SPACE TRAVELERS
AND
THE GENESIS OF
THE HUMAN FORM

Evidence of Intelligent Contact in the Solar System

by Joan d'Arc

THE BOOK TREE
ESCONDIDO, CALIFORNIA

**Space Travelers and the Genesis of the Human Form:
Evidence of Intelligent Contact in the Solar System
ISBN 1-58509-127-8**

VOLUME TWO IN THIS SERIES:

Phenomenal World
ISBN 1-58509-128-6

LAYOUT AND DESIGN
Tedd St. Rain

COVER ART:
Pakal by Charles William Johnson & Jorge Luna, ©1995-1999 by Charles William Johnson. Used with permission from the book Earth/matriX: Science in Ancient Artwork. P.O. Box 231126, New Orleans, LA 70183-1126, (504) 733-9291 Tel/Fax, or visit us at www.earthmatrix.com.

Printed on Acid-Free Paper

Dedicated to the memory of my father,
Frederick,
and his peculiar paranoid logic.

Hey Dad, it works!

Acknowledgements

Grateful acknowledgements to the iconoclastic living "giants" on whose shoulders I attempt to stand, other amnesia victims who have been bitten on the ass by forgotten human knowledge, among them: Charles Fort, Zecharia Sitchin, Richard Hoagland, Robert Temple, Phillip Johnson, Mary Midgely, Trevor Palmer, Michael Cremo, Richard Thompson, D.S. Allen, J.B. Delair, and the list could go on; but there would be no list without the giant on whose shoulders they stand: Immanuel Velikovsky.

Special gratitude to Robert Guffey for reading my first manuscript and giving me encouragement to get beyond the lame "literary agent" attitude that the book was "too dense" for mainstream consumption. Thanks for the detailed feedback and letters of support when I almost gave up. Gratis to my publisher, The Book Tree, for giving this work a fighting chance.

Kind gratitude to Charles William Johnson, for so graciously allowing me to use his beautiful artwork to grace the cover of this book, and for allowing me to quote material from his *Earth/matriX: Science in Ancient Artwork* website (www.earthmatrix.com). Thanks to the Shadow Robot Company Ltd. for kind permission to publish the Shadow Robot Walker (www.shadow.org.uk) and/or (www.androidworld.com) and to The Cog Shop at MIT (www.ai.mit.edu.projects.cog/) for persmission to picture their creation. Thanks to Hoagland's website, www.enterprisemission.com, for continually blowing my mind.

Volume two in this series, *Phenomenal World*, (ISBN 1-58509-128-6, available from The Book Tree) picks up where this book leaves off.

Table of Contents

Introduction

There is practically no person on this *terra firma* who does not have an opinion regarding the question of the existence of extraterrestrial intelligent life (ETI). If you are a believer in ETI, you have no doubt been confronted with the few standard arguments covered in this book that are pitched by most skeptics. But are they logical and internally consistent? Or are they based on mistaken assumptions, government-media hogwash, and outmoded scientific concepts? Even if you are a skeptic, you may want to explore the logical grounds upon which your staunch protest against the existence of ETI is founded. Therefore, *Space Travelers and the Genesis of the Human Form* is for everyone on the planet.

One of the most persistent arguments against the reality of the Visitor Experience (alien visitation or abduction) is based on the presumption that if the incrementally accidental and unguided evolution of the humanoid form occurred on Earth, it is nearly mathematically impossible that it could have occurred on another planet. The assumptions of Darwinian evolution presuppose the humanoid form to be entirely Earth-based. Remarkably, the assumption of the accidental evolution of mankind from the great ape lineage is always overlooked as the problematical factor in the analysis. *Space Travelers and the Genesis of the Human Form* illustrates that Darwinian evolution is actually not an empirically predictable or testable scientific paradigm, but is a highly touted philosophy of Western materialism.

Space Travelers and the Genesis of the Human Form begins with a discussion of the anthropic principle, which asserts that the evolution of "conscious observers" requires a minimum of a billion years. The Space Travel Argument Against the Existence of ETI is based on this billion year requirement. Therefore, this anthropic argument assumes that if extraterrestrial intelligence existed, it would have been here by now. Of course, the only "evidence" for ETI contact which is taken into account is "radio signals." Any other potential evidence is denied. A survey of ancient anthropomorphic artifacts on Mars and the Moon illustrates that evidence of so-called "Game Wardens" in our own solar system actually exists.

The Space Travel Argument Against the Existence of ETI will be shown to be dependent on three factors: (1) the persistent imposition of Earth-centered technological constraints (specifically, rocket technology and radio signals) implying an anthropocentric 'you can't get here from there' attitude; (2) mathematical logic deduced from the faulty linear notions of Darwinian evolution, which only serve to put the 'cart before the horse'; and (3) a circular and untestable hypothesis which essentially states 'they aren't here because they aren't here.'

Space Travelers and the Genesis of the Human Form also outlines a new paradigm which may account for the appearance of similar life forms on disparate worlds. The Extraterrestrial Seeding hypothesis is based on the prolific written accounts of the "first" known Earth civilization, the ancient Sumerians, as well as the vast Vedic literature and early Greek literature, and the oral traditions of aboriginal peoples, including the Dogon tribe of Mali. These sources are typically viewed from a neo-Freudian angle as a "spontaneous projection" from the mind of "archaic" man. Conversely, the dramas of these mythogenic sources can be seen as an exogamously borne pre-history of the planet. These mythological dramas are reviewed in the context of ancient markers—human artifacts—left on planetary surfaces in our own solar system, including Earth, the Moon, and Mars.

The current consensus reality appears to hold out for a Science of Exclusionism: isolation of the Earth. From Charles Fort's *The Book of the Damned*, this book borrows the following proposition:

> This book is an assemblage of data of external relations of this earth. We take the position that our data have been damned, upon no consideration for individual merits or demerits, but in conformity with a general attempt to hold out for isolation of this earth.

> Having attempted to systematize, by ignoring externality to the greatest degree possible, the notion of things dropping in upon this earth, from externality, is as unsettling and as unwelcome to Science as tin horns blowing in upon a musician's relatively symmetric composition, or flies alighting upon a painter's attempted harmony...

> So, our pseudo-standard is Inclusionism, and, if datum be a correlate to a more widely inclusive outlook as to this earth and its externality and relations with externality, its harmony and Inclusionism admits it.

> I think of as many different kinds of visitors to this earth as there are visitors to New York, to a jail, or to a church...

The Science of Exclusionism is the rule by which we acknowledge "truth," but, as Charles Fort asserts, "science is a Turtle that says that its own shell encloses all things." Materialist doctrines describe the world as a closed system: isolated humankind in an isolated consciousness on an isolated oasis. Nothing enters this closed world from the Outside.

Materialist cause and effect rhetoric, particularly Freudian and Darwinian paradigms, serve to keep Earth humans earthbound, since they keep us from potentially adding up 2+2 with regard to our true ancestry from the "sky" rather than from the "water." In this peculiar insular point of view, we have only just recently accepted the fact that things might fall onto the Earth from the Outside. Yet, the current scientific paradigm allows no exogamous contact with anything resembling "human" consciousness. In our egoistic awareness, we continue to label all consciousness as derivative of ourselves, and assume ourselves to be the first "conscious observers." All theories are then based on this presumption.

As this book will illustrate, the Space Travel Argument Against ETI is a circular, self-contained, and, therefore, tautological, argument. *Space Travelers and the Genesis of the Human Form* will attempt to tear down the walls of this insulated system. Consider this book an experiment in Exogamy; a hold out for a Science of Inclusionism: a science that *encompasses* the external, the existential, the extraordinary, the extrasensory, the exogenous, the excluded, and the extraterrestrial.

Chapter One

Robot Probes and Indefinite Survival:

Von Neumann Probes, O'Neill Colonies & The Zoo Hypothesis

The advancement of knowledge should be the only objective of scientific observation and experimentation, rather than the propagation of commonly-held belief systems.

Tom van Flandern

quoted from *The Anomalist*

In its "Astrobiology" web site, NASA claims an interest in the search for past or present life on Mars and Europa, two solar system bodies which once supported water. Among the goals of Astrobiology are the following: to determine what makes a planet habitable and how common these worlds are in the Universe, how to recognize the signature of life on other worlds, how life arose on the Earth, how matter organizes into living systems, how the terrestrial biosphere has co-evolved with the Earth, and what are the limits for the future of life on Earth.

Part of understanding how life arose on Earth includes the possibility that "it arrived at Earth from elsewhere." They state: "terrestrial life is the only form of life that we know, and it appears to have arisen from a common ancestor." The newfound discipline of Astrobiology is apparently open to the concession that this ancestor (i.e. microbe) arrived in a meteor but, within the current Western cosmology, meteors and comets and such are the only celestial flying objects that might harbor signs of life. The Astrobiology web site is clear in stating that it has not ascertained whether "life from one world can establish an evolutionary trajectory on another," but it is considered one of their goals to understand such natural migratory processes. Can the "seed" of life grow into a civilization of intelligent humanoids on its own accord?

One of the goals of Astrobiology is " the search for extraterrestrial life and the potential to engineer new life forms adapted to live on other worlds." According to NASA's Astrobiology Roadmap, Objective 16 is to "understand the human-directed processes by which life can migrate from one world to another." They write:

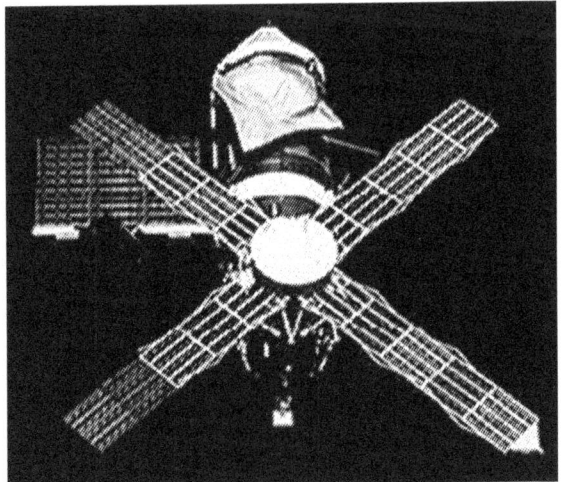

Mankind attemps to colonize space with Skylab in the 1970s.

> For the first time in human history, we can intentionally move life beyond our home planet. As a result, humanity is entering a new evolutionary territory—space—in a manner analogous to the first sea creature crawling out onto the land, with the attendant requirement for supporting technology. This time,

9

however, we are able to document this evolutionary trajectory with the tools of modern molecular biology and to engineer artificial ecologies that may be necessary for evolutionary success in this new environment.

As we will soon see, the Space Travel Argument Against the Existence of Extraterrestrial Intelligence suggests that the ultimate survival of a technological civilization lies in dispersing its genetic material out into the Universe; seeding itself into other unoccupied niches. While NASA is obviously following the dictates of this theoretical stance in its study of "human directed" migration of life from one world to another, it does not recognize (at least openly) that such could have already occurred. Neither does it recognize the anthropomorphic evidence which would indicate that such a migratory pattern has been ongoing within our own solar system and on our own planet.

As we can safely deduce from the above discussion, NASA *is* interested in "life" in the solar system, but it is clearly interested in microbial life, not intelligent life (i.e. humanoid). The evolutionary paradigm itself is the controlling factor, and is the reason for major cover-ups of potentially man-made artifacts on the Moon and Mars. It is obvious that the discovery of humanoid intelligence would upset scientism's apple cart: evolution. But have we put the cart before the horse? This book will explore these ideas from as many angles as possible.

Prominent evolutionist Theodosius Dobhzhansky has posed a hypothetical situation wherein by some "utterly unlikely chance" there is another planet on which there arose animals and other living forms similar to those of the Eocene period. He asks, would "manlike creatures" also develop on this planet? He answers that the probability that mutations and genetic alterations in the 50,000 genes required for this development to occur, over a span of 55 million years, would be "virtually zero." Even small deviations in the "sequence of changes," he argues could throw evolution "off the track to humankind." He also points out that deviations in climate from that of the earth's climatic history could also "derail human evolution."

Evolutionist George Gaylord Simpson has defined a humanoid as "a natural, living organism with intelligence comparable to man's in quantity and quality, with the possibility of rational communication with us." In his book *Alien Identities*, Richard Thompson outlines Simpson's argument that human evolution is contingent on a "vast number of special circumstances" which are extremely unlikely on another planet, including chemical, environmental, and

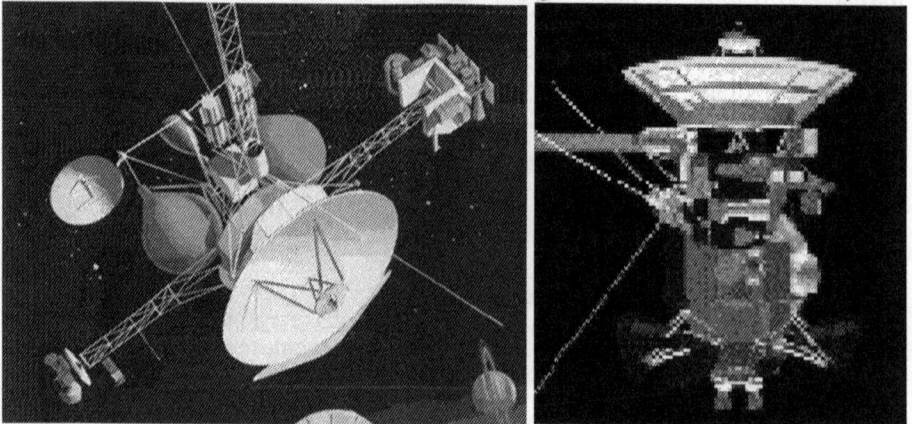

Launched on October 15, 1997, the Cassini space craft includes the orbiter and the Huygens Titan probe. It is the largest, heaviest, and most complex interplanetary space craft ever built. The power source for Cassini is radioactive uranium. Instruments aboard the probe will analyze the atmosphere of Saturn. Illustrations commissioned by Jet Propulsion Laboratory, 1988.

mutational circumstances. Simpson concludes that it is "extremely unlikely that anything enough like us for real communication of thought exists anywhere in our accessible Universe."

The "Fermi Paradox" states that Earth contains the only advanced civilization in the galaxy, "since if there were others we would know about them." Conversely, other scientists have estimated that there could be billions of advanced technological civilizations in the galaxy. Neither of these speculations, or any in between, have been proven by direct evidence and must remain

The Galileo Descent Module as seen in 1983, descended into the Jovian atmosphere.

matters of opinion. Nonetheless, in the 1992 *Journal of the British Interplanetary Society*, British scientist, E.J. Coffey argued:

> The evolutionary conclusion that humanoid intelligence elsewhere is improbable is not due to any anthropomorphic bias, but because of the deep understanding that evolution has no real goal other than adapting creatures to specific local environments. Neither we, nor our mode of intelligence, are the high point of evolution. The pathways of evolution are too circuitous for that ever to be the case. As you see, the scientific bias precludes space travel, since it sees evolution as adaptation to specific local environments.

Coffey believes that by aiming its "very costly radio telescopes at the stars," NASA and its supporters hold a "religious conviction" for the search for intelligence in the Cosmos. Actually, the SETI project has been careful to make the distinction that it is searching for extraterrestrial "intelligence" which, as ancient historian Zecharia Sitchin has quipped, could just mean "smart rocks." It has never suggested it is searching for, or expects to find evidence for, humanoid intelligence. Sitchin has queried why an advanced civilization would be using something as primitive as radio signals. The notion of looking for coherent radio signals within the background of radiation in space, he has suggested, is "not just a waste of taxpayer money but is really an effort to say 'don't look *right here*, look there,' because there you won't find anything."

The power to define what constitutes evidence of extraterrestrial intelligence is, as we shall see, an important facet of the tautological argument against the existence of extraterrestrial intelligence.

The Space Travel Argument

The Space Travel Argument Against the Existence of Extraterrestrial Intelligent Life (Space Travel Argument), explained in detail by Barrow & Tipler in *The Anthropic Cosmological Principle*, begins with the telling statement that those who propose that extraterrestrial intelligence (ETI) probably exists in the Universe tend to be physicists and astronomers, and those who tend to argue *against* the probability for ETI are more likely to be evolutionary biologists. Thus, the crux of the Space Travel Argument is founded on the theory of Darwinian natural selection, the adaptation of creatures to their local habitats.

The Anthropic Principle is based on a biological argument: the minimum time required for the evolution of "intelligent observers." A billion years is required for the evolution of intelligence; therefore, a star must have been stable for at least that long. The Anthropic Timescale Argument allows that the types of processes allowed in the Universe must be of such an age that "slow evolutionary processes will have had time to produce intelligent beings from non-living matter." (Barrow, 159)

The Space Travel Argument is based on the Anthropic Principle: the mathematical assumption that a "communicating species" would evolve in less than 5 billion years and would eventually begin interstellar travel. This argument contends that "since 1 billion years is quite short in comparison with the age of the Galaxy, it follows *from the absence of ETI* in our Solar System that such space-travelling ETI apparently *do not exist*, and *have never existed* in our Galaxy." The authors note that this assumption is *logically inferred from observed evidence*, and from astrophysical observations and theories. It should be stressed that Barrow & Tipler explicitly assert that *absence of evidence is evidence of absence*.

The Space Travel Argument states specifically that "the contemporary advocates for the existence of extraterrestrial intelligent life seem to be primarily astronomers and physicists, such as Sagan, Drake, and Morrison, while most leading experts in evolutionary biology, for instance Dobzhansky, Simpson, Francois, Ayala et al. and Mayr, contend that the Earth is probably unique in harboring intelligence." The Space Travel Argument contends that the "probability of the evolution of creatures with the technological capability of interstellar communication within five billion years after the development of life on an Earthlike planet is less than 10^{-10}" and, therefore, they deduce, "it is very likely that we are the only intelligent species now existing in our Galaxy." The Space Travel Argument also asserts that if an advanced interstellar civilization did exist and possessed the technology for interstellar communication, "they would also have developed interstellar travel and thus would already be present in our Solar System. Since they are *not here*, (footnotes 14, 15) this implies that they do not exist." (Barrow, 576)

Whether or not one believes that extraterrestrial visitors have ever traversed our Solar System, the logical continuum of the above argument must be addressed. Incredulously, the footnotes following the statement 'they are not here' reference the 1974 book *UFO's Explained* written by the most popular UFO debunker, Philip Klass. Thus, their *proof*—point blank—that interstellar visitors are *not here* is the obviously highly regarded opinion of a member of The Committee for the Scientific Investigation of Claims of the Paranormal (CSICOP), who has been labeled the most negative of the UFO debunkers. As quoted in *At the Threshold*, UFO researcher, Dr. David Jacobs, has summed up the attitude behind the culture of debunking:

> Debunkers are not open minded skeptics. The combination of ignorance of the subject, a messianic sense of defending science from the forces of superstition, an ego-charged idea that they know the answer to whatever UFO problem is being discussed, and a streak of mean-spiritedness are the necessary ingredients for debunking.

Barrow & Tipler fail to further qualify the deduction that *they are not here*, and only mention in passing that there might be a problem with this assumption. The authors seem to be espousing a tautological argument which essentially states that extraterrestrials are not here and never have been here, because they are not here and never have been here. As we shall see, the Space Travel Argument is also an anthropocentric argument founded on the proposition that human intelligence *evolved* as a purely local phenomenon on an outback planet and, further, that this accidental (local) event has no Universal (i.e. external) relationship.

The Space Travel Argument contends that an intelligent species with the technology for interstellar communication would also develop the technology for interstellar travel which is *at least comparable to our present-day technolo -*

gy, and particularly rocketry, leading automatically "to the exploration and/or colonization of our Galaxy in less than 300 million years." This deduction is based on the "Principle of Mediocrity" which states that "our evolution is typical of the evolution of any species on any planet capable of supporting life." In addition to the likelihood of possessing rocket technology *comparable to our own*, the authors posit that it seems probable that such a species would also possess sophisticated computer technology. The Space Travel Argument unequivocally equates the concept of *survival of the fittest* to the technological ability to disperse human DNA into the Galaxy utilizing a theoretical computer called a von Neumann probe.

Von Neumann Probes

Carl Sagan has proposed that "communication with extraterrestrial intelligence will require, if our experience in radio astronomy is any guide, computer actuated machines with abilities approaching what we might call intelligence." The Space Travel Argument, therefore, assumes that "any species engaging in interstellar communication will have a computer technology not only comparable to our present-day technology, but which is comparable to the level of technology which we know is *possible*, which we are now spending billions of dollars a year to develop, and which a majority of computer experts believe we will actually possess within a century." Specifically, such an interstellar species will have "a self-replicating universal constructor with intelligence comparable to the human level." Such a machine, combined with present-day rocket technology, would make it possible to explore the Galaxy in less than 300 million years. They further assert that it is "a deficiency in present-day computer technology, not rocket technology, which prevents us from beginning the exploration of the Galaxy tomorrow."

The Space Travel Argument also contends that "the ultimate survival of a technological civilization, and indeed the survival of the biosphere in some form, requires the eventual expansion of the civilization into interstellar space." They suggest that the "resources available in uninhabited stellar systems cannot be utilized for any human purpose unless a space vehicle is first sent;" therefore, "optimal exploration strategy must utilize the materials available in other stellar systems as far as possible." Utilizing these otherwise useless interstellar resources will cover the costs of space exploration. (Barrow, 578)

The crux of the Space Travel Argument is the aforementioned "self-reproducing universal constructor;" a computerized machine "capable of making any device, given the construction materials and a construction program." Such a machine, called a *von Neumann probe* after the physicist who has shown that such a machine is theoretically possible, is by definition capable of making a copy of itself. A probe sent to another stellar system would include a self-replicating universal constructor, with human-level intelligence, capable of self-repair and self-programming, with an engine for slowing down once the stellar system is reached, and an electric or solar propulsion system for travel within the system.

The universal constructor would be instructed to carry out a search for construction materials and would set out making copies of itself, and the rocket engines and other devices needed, from the interstellar resources. The authors contend that such materials should be in abundance in the vicinity of most stars in the form of "meteors, asteroids, comets and other debris from the formation of the stellar system." For instance, they explain, many asteroids are made up largely of nickel-iron and some contain large amounts of hydrocarbons. Copies of the probe would be launched at the nearest stars, and the process would be repeated on those stars. The probes would then be programmed to explore the stellar system and relay the information back to the original probe.

The O'Neill Colony

The authors explain that the von Neumann probe can also be used to colonize the stellar system, even if there are no planets in the system, by programming it to turn some of the available space debris into an *O'Neill Colony*: a space station which contains a self-sustaining human colony. The authors explain that "all the information needed to manufacture a human being is contained in the genes of a single human cell." It is doubtful that frozen cells would remain viable over the long period of time required to cross interstellar distances via space probe. Thus, a theoretical scenario which overcomes this problem is explained (Barrow, 580):

> If an intelligent extraterrestrial species possessed the knowledge to synthesize a living cell—and some biologists claim the human race could develop such knowledge within 30 years—they could program a von Neumann probe to synthesize a fertilized egg-cell of their species. If they also possessed artificial womb technology—and such technology is in the beginning stages of being developed on Earth—then they could program the von Neumann probe to synthesize members of their species in the other stellar system ... These beings could be raised to adulthood in the O'Neill Colony by robots also manufactured by the von Neumann probe, after which these beings would be free to develop their own civilization in the other stellar system.

It is thought that the information required to synthesize an egg cell would tax the memory of the probe, so the information would be transmitted by microwave to the probe once it constructed additional storage capacity. Theoretically, a von Neumann probe cannot become obsolete, since it is a universal constructor which can be updated via radio with instructions for creating new technological devices.

The Space Travel Argument states that it "seems reasonable to assume that any intelligent species would develop at least the rocket technology capable of a one-way trip with deceleration at the other stellar system." They assert that the transport of a von Neumann probe to an interstellar system could be accomplished with present day rocket technology, and that nuclear power is not necessary. The authors later concede that "restricting consideration to *present-day rocket technology* is probably too conservative," and an advanced intelligent species would "eventually *develop rocket technology* at least to the limit which we regard as *technically feasible today*."

The most restrictive problem with regard to space exploration, however, is the cost of rocket fuel. For example, the major cost of the Orion Project study was the cost proposed for deuterium fuel, which amounted to the present GNP of the United States. The cost for helium-3 in the British study for the proposed Project Daedalus interstellar probe amounted to about 100 times the U.S. GNP. Due to the cost of rocket fuels, the feasibility of our sending out such probes using rocket fuel, they admit, is "far beyond the means of present day civilization." The problem of cost restrictions is solved by the theoretical von Neumann probe, since probes construct themselves from available space debris, and the initial probe can be launched by chemical rocket. The cost of exploring the Galaxy with a low-velocity von Neumann probe, they ascertain, would be about 30 billion dollars, approximately the cost of the Apollo program. (Barrow, 583)

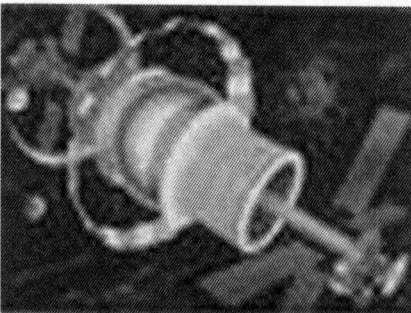

The Bernal Sphere, an artistic study of a habitable space colony by Rick Guidice, 1976. NASA Photo Archives #AC76-0965

A Biological Machine

The Space Travel Argument contends that "a human being is a universal constructor specialized to perform on the surface of the Earth." After making this statement which, in effect, reduces the human to a biological machine, Barrow & Tipler go on to assert that von Neumann probes should have the same rights as human beings. They launch a peculiar discussion of human rights and how those can be extended to a von Neumann probe, which is after all an "intelligent being in its own right, only made of metal rather than flesh and blood." They contend that "arguments against considering intelligent computers to be persons and against giving them human rights have precise parallels in the nineteenth-century arguments against giving blacks and women full human rights." They appear to be hopeful that in the future "von Neumann probes would be recognized as intelligent fellow beings, beings which are the heirs to civilization of the naturally evolved species that invented them." After all, they contend, the "naturally evolved species and all of its naturally evolved descendants must inevitably become extinct ... but ... a civilization with machine descendants could continue indefinitely." (Barrow, 595)

Newman and Sagan argue that an advanced technological civilization would not necessarily be motivated to explore and/or colonize the Galaxy. Conversely, Barrow & Tipler find it difficult to imagine a "plausible scenario" whereby an intelligent species would have the technology and expertise to begin interstellar travel, but would decide not to deploy it. They contend that once probes have been launched, they will explore the Galaxy automatically, whether or not the civilization that launched them continued to exist. They surmise that, if such intelligent beings with the desire for interstellar communication did exist they would be quite capable of developing probe technology, however, they *might choose not to use it*. The Space Travel Argument concludes that there is no good reason for believing this to be true.

The following statements are especially pertinent here. One possible reason for a decision not to build probes, the authors concede, is the fear of *losing control* of them. They concede that it is possible that the program which keeps the probe under the control of the intelligence species could be accidentally omitted during the reproduction process, with the result that the copy *goes into business for itself*. They indicate that this problem can be surmounted by keeping the control program so integrated with the whole that it would fail to reproduce at all if something went wrong with that part of the program. They point out that this is analogous to constraints imposed on cells used in recombinant DNAtechnology.

Following is a discussion of the points made by the Space Travel Argument which are pertinent to the argument at hand, that is, whether its conclusion that Earth humans are alone in the Universe is based on logical grounds.

Survival of the Fittest

It has been noted that scientists who tend to argue *against* the probability for ETI are more likely to be evolutionary biologists, therefore, the Space Travel Argument Against the Existence of ETI is based on the *theory* of Darwinian evolution. The problem with this assumption is that Darwinian theory is not a testable scientific hypothesis. It is a tautology: a self-contained system of circular proofs which are always true in a self-contained system of circular proofs. We will look in detail at Darwin's theory of evolution in a later chapter. In the meantime, think on this: a circular dilemma confounds the popular use of the evolution argument against the co-existence of the humanoid form in the Cosmos, since we do not actually *know* that we are the only humanoids in the Universe, nor do we *know* the genesis of the humanoid form. We are simply extrapolating an earthbound premise from an Earth-centric theory.

The Space Travel Argument essentially deduces that human evolution is typical of the evolution of any species on any planet capable of supporting life. Furthermore, according to this argument, the most coveted "talent" which us Darwinian-evolved humans want to exhibit more than anything else is something called species survival. In the human species, ultimately "survival" means spreading your DNA out into the stars in case of a major planetary catastrophe. Perusal of NASA's "Astrobiology Roadmap" web site will illustrate that NASA's objectives are none other than to seed other planets in our solar system with life forms from Earth, including humans. The future extensions of Astrobiology's Objective 16 include the following: "engineer closed and open environments as prototypes for human exploration of other planets, test such systems in analog environments on Earth and in space, place candidate ecosystems on extraterrestrial surfaces and document their evolution, establish permanent colonies of humans and other organisms in space and on another planetary surface, engineer life for survival, adaptation and evolution beyond Earth."

Assuming that a behavior pattern typical of *Homo sapiens* would be adopted by any intelligent species, Barrow & Tipler maintain that an advanced civilization would launch something like von Neumann probes or colonization ventures of some sort. They contend that "all living things have a dispersal phase, in which they tend to expand into new environments," and that these are "generally carried out to the limit allowed by their genetic constitution." In an intelligent species, they write, "this limit would be imposed by the level of technology," and further they propose, using explicit Darwinian lingo, that "those groups which do not exhibit this behavior would be *selected against*." Therefore, it is to be expected that at least some groups of interstellar species would attempt expansion into the Galaxy. The launching of something like a von Neumann probe would increase the probability that the civilization will survive the death of its star, nuclear war, or other catastrophes. In order to accomplish "indefinite survival," the authors suggest, space colonization would not need to be of an 'imperialist' nature, but could consist of the construction of space stations revolving around stars.

The Space Travel Argument makes the Earth-centric presumption that "survival" alone, and not any moral ground, is the fundamental impetus for the behavior of any civilization. This survival of the fittest thesis supposes that a civilization which chooses not to engage in colonization behavior would be, naturally, "selected against." The authors present the anthropomorphic premise that they find it "difficult to construct a plausible scenario" which would explain why an ETI civilization would have the technology for interstellar travel and *would not engage in it*. Thus, this position is based on the presumption that if they *can* do it, and they are *not doing it*, then it follows that they do not exist.

The Space Travel Argument is rigidly anthropocentric and circuitous. We cannot extrapolate the behavior of interplanetary citizens from Darwinian-based observations of our own behavior. Darwinian evolution cannot be used as a framework from which to argue against the co-existence of the humanoid form in the Cosmos at large, since this assumption could likely be placing 'the cart before the horse'.

If the question we are asking is whether Earth humans are part of an intelligently designed family dispersed in the Cosmos at large, or whether we are alone in an immense deaf and dumb Universe with only "smart rocks" as neighbors, it is illogical to begin with any presumption at all.

From the perspective of Darwinian evolution, if ETI exists at all, it absolutely cannot have the same form that we happen to have, because it will crush the scientific premise of our purely accidental climb out of the ponds of our local habitat Earth. How could creatures on a distant clod of Earth accidentally repeat the same incremental and circuitous climb while being victims of an entirely different "lottery"? They can't. Then perhaps we need to address our *a priori* premise that the emergence of the humanoid form is a purely local phenomenon.

Earthbound Thinking

It is ludicrous that we should assume any Earth technology to be a cosmic constant. Why does the Space Travel Argument limit the universal possibilities of interstellar travel to rocket technology, and of interstellar communication to radio signals? Why does it limit the universal possibilities of space travel to "intelligence comparable to the human level?" Is it logical to assume that the "evolution" of science and technology necessarily proceeds along the same course on all planetary systems? And, even if it were the case, is the science behind "rocket technology" necessarily the apex of any advanced technological civilization ready for countdown to space exploration?

While surely being aware that an electromagnetic propulsion system would be necessary for distant space journeys, the Space Travel Argument emphatically presumes that ETI technology would be based on or comparable to the science behind "rocket engines" and "radio signals." With regard to radio signals, Barrow & Tipler admit that space probe contact has various advantages over using radio waves; one being the problem of guessing the radio frequency used by the other species. Additionally, by concluding that a "deficiency in present-day computer technology, not rocket technology" is what prevents Earthlings from space exploration, they are admitting that space probes are superior to rocket technology as a way to move DNA from one place to another. On one hand the authors state that an advanced intelligent species would eventually develop rocket technology at least to the limit which we regard as *technically feasible today*, while they conversely argue that the cost of rocket fuels puts deep space exploration and colonization far beyond the means of present day civilization.

You don't have to be a rocket scientist to realize that we don't see the rocket ships of extraterrestrial travelers in our skies because rocket technology is an unfeasible, in fact, impossible, way to explore galaxies. Some other type of system, backed up by an extraordinarily different type of science and technology, is needed. And since we are so incredibly far from achieving the ability to traverse the galaxy with our own technology, we conclude that because *we* can't get *there* from *here, they* can't get *here* from *there*.

Thus, the term "technically feasible" seems to outline the boundaries of a paradigm to which we are technologically leashed; a technology which serves to limit the collective vision of the society at large. Philip Corso writes in *The Day After Roswell*, that the problem of long-term space travel has not been solved partially because we continue to utilize conventional means of propulsion, putting astronauts under enormous duress during takeoff. He asserts that humans will not reach destinations beyond the solar system without a "radically different" form of propulsion. Paradoxically, many have suggested that, under cover of national security, top U.S. defense contractors are secretly developing such technology.

Alternative theories such as electromagnetic antigravity propulsion were well known in the 1920s; prototypes were being developed by Paul Biefield and Townsend Brown at the California Institute for Advanced Studies, but they never got far off the ground. Neither did the German prototypes which originated from Tesla's work in Russia, although allied air forces assumed they had when flying saucers first appeared in the skies during World War II. But those flying contraptions were of a different order entirely. As Corso writes, the effort to develop true antigravity aircraft "never came to fruition among conventional aircraft manufacturers because gasoline, jet and rocket engines provided a perfectly good weapons technology."

Therein may lie the crux of the matter. Barrow & Tipler make the assumption that a "behavior pattern typical of *Homo sapiens* would be adopted by any intelligent species." What, essentially, is the behavior pattern of the species *sapiens sapiens*? Who would argue that what we consider "progress" essentially pays for itself first by being based on "a perfectly good weapons technology"? War "behavior" is the behavior pattern typical of Earth humans; but we should *not* assume it is typical of "any intelligent species." On the contrary, nuclear physicist Stanton Friedman suggests in his book *Top Secret/Majic*, that extraterrestrials are here "to make sure that our brand of 'friendship' is not visited upon other civilizations in the neighborhood."

The Ancestral Earth Boundary

In his book *The Sirius Mystery*, Robert Temple has also addressed the question of our readiness to join the galactic federation. He speculates that the only societies that might be carrying on an interstellar dialogue of any kind are the "magical societies;" that is, societies so advanced that we are but "emerging primitives" in comparison. Such societies would most certainly have standard procedures for dealing with primitives, and may even have "commenced their operations with the long range intent of bringing us into their club." He asserts, however, that "just as no London gentlemen's club wishes to have a savage in a g-string waving his spear and poisoned arrows about in the member's lounge, so the interstellar club is unlikely to plug us straight into the circuits as a fully-fledged member."

It is apparent that a technology based primarily on killing, and secondarily on consumerism, has brought us down the wrong path with regard to "survival." It has and continues to stunt our growth as a truly intelligent species. It has essentially kept us primitive. Yet, the Space Travel Argument propagates the notion that this technology—basically only suitable for bonking each other over the head, or making threats to the same effect—is the apex of scientific prowess. This is an extremely anthropocentric attitude. We're so primitive that we don't know we're primitive.

As Friedman asserts, space travel first requires that a civilization learn to be at peace with its neighbors. In terms of *its own survival* then, any advanced civilization that gets a load of us will surely have reason for serious concern. As Friedman speculates in *Top Secret/Majic*, an advanced society would probably pay attention to the more primitive societies in the neighborhood, and would pay "especially close attention to those showing signs of venturing beyond ancestral boundaries." Friedman surmises that it would be to their advantage to keep us from exploring our solar system or beyond, since they see us as essentially a hostile race in the toddler stage of growth.

Ironically, while we might see space travel as survivalist behavior, an advanced civilization may see their own survival threatened by such behavior and, thereby, would do everything in their employ to keep the Earth children earthbound. Ancient writings indicate that Earth humans are bound to the planet Earth. As Psalm 115 states, "The heavens are the Lords, but the Earth has he given to the children of men." This book will explore the idea that, at least in our material lives, we are bound to this *terra firma*.

In his book *The Sirius Mystery*, Robert Temple discusses an ancient Greek reference to the babysitting of the human race, called *The Virgin of the World,* which describes the hierarchical principle of lower and higher beings in the Universe. This treatise essentially describes a situation wherein Earthly mortals are presided over by higher beings who interfere in Earth's affairs when "things become hopeless." The *Virgin of the World* also describes a personage called Hermes (or Anubis) who "seems to represent a race of beings who taught Earthly mankind the arts of civilization after which, 'with charge unto his kinsmen of the Gods to keep sure watch, he mounted to the Stars'." This ancient treatise suggests that "mankind have been a troublesome lot requiring scrutiny and, at rare intervals of crisis, intervention."

James Deardoff of the Department of Atmospheric Sciences at Oregon State University has theorized about the possibility that the Earth is under quarantine. He has written that our "lack of detection" of ETI life forms could just as well be an indication that an "embargo" of sorts is in place. Deardoff has suggested that such an embargo might be aimed at our premature and sudden discovery of them, and our subsequent panic, which might end up in a nuclear exchange. He speculates that any sudden lifting of the embargo in a manner obvious to the public would cause societal chaos. Therefore, according to Jim Marrs in *Alien Agenda*, the lifting of an embargo must be designed to allow gradual disclosure of the alien message in order to allow dissemination of information over a long period of time.

As Deardoff has also noted, any radio communications received by ETI would likely be heavily censored by government agencies. So, how can we accept "absence" of radio signals as "evidence of absence"? In addition, as Stanton Friedman has pointed out, those awaiting radio signals from distant stars often presume visitors would be coming from galaxies millions of light years away, and that they would be using "the dumb old chemical rockets we are stuck with now." On the contrary, if they were using such technology, we certainly should not bother baking a cake.

Furthermore, as Friedman has pointed out in *Top Secret/Majic*, they would *not* necessarily be coming from millions of light years away, since "the galactic neighborhood is not as big as some researchers make it out to be." There are about 1,000 stars within our local galactic neighborhood, or within 54 light years of Earth: "a mere walk down the block by galactic standards." Among those 1,000 stars, Friedman contends, about 46 of them are very similar to our sun. Most astronomers agree that sun-like stars are likely to have planets. In particular, the stars Zeta 1 and Zeta 2 Reticuli in the constellation Reticulum are only 37 light years away from Earth. These stars are about a billion years older than our Sun. He writes: "a civilization that had a billion-year head start on us will certainly know things that we can't even dream of."

The Zoo Hypothesis

Barrow & Tipler present the scenario that if an ETI civilization was reluctant to contaminate the culture of another species with its own culture, it may decide not to attempt radio contact. The authors suppose that "with probes it would be possible to study an alien species without it becoming aware of the species which is studying it." Thus, probes could maintain contact and control via secrecy. The Zoo Hypothesis suggests that the messenger probes of an ETI species "have been here for a long time but have decided not to make their presence known."

In this regard, Temple has written that if interstellar travel has become possible for a few or more extraterrestrial societies, they have almost certainly visited Earth in its lengthy history as a planet. There is no doubt, he writes, that "our distant ancestors the cave men would have been observed by extraterrestrial probes, who would have made a note that something was happening on this planet—slowly happening, but nevertheless still happening." He surmises that if this were so it would certainly have "impacted mankind" and would have "been incorporated somehow into his traditions."

The Space Travel Argument states that a human being is essentially "a universal constructor specialized to perform on the surface of the Earth." There is no sound argument which can be asserted against the existence of an ETI probe in the past and it would, theoretically, explain the genesis of life forms on Earth. Explaining that "all the information needed to manufacture a human being is contained in the genes of a single human cell," Barrow & Tipler assert that an advanced civilization could "easily synthesize members of their species in the other stellar system." They suppose that those species would then be "free to develop their own civilization in the other stellar system." They go on to admit that it would be possible to study an alien species using such probes without that society becoming aware they were being studied.

To take this seemingly preposterous scenario another step further, could the entire Earth biosphere also be bio-engineered? Along these lines, H.V. Ditfurth has argued in *Origins of Life* (p. 51) that "it was not the case that life on Earth was dependent on only twenty quite specific amino acids (and countless other molecular elements) from among hundreds of possible ones." He writes that the limits of the supposed "right conditions" should not be thought of as overly narrow. In fact, the way that the molecular pattern of the cosmos matches the Earth's biological one might suggest something else entirely. Ditfurth explains that when the Earth's history began there was no specific *need* for organisms to live on it. In other words, he asserts, the reason why these particular 20 amino acids are present in the cells of all Earth organisms is not that there was no other way for life to get started, but, rather, Earth organisms were constructed with these particular building blocks simply because "they were present in great abundance."

Might we state this another way? Earth's habitat and life forms are constructed from a finite collection of material that was just hanging around in the area doing nothing. Recall Barrow & Tipler's "optimal exploration strategy" which must utilize "otherwise useless interstellar resources." Seen from this perspective, could all of Earth's biological forms be the result of a remote engineering job performed by the universal constructors of an extraterrestrial race?

Based on the Space Travel Argument, the speculation that (1) intelligent life has the tendency to expand into and occupy space, and (2) such expansions will utilize raw materials in the vicinity of a planetary system for the building of space colonies, Papagiannis and others have argued that if ETs are in our solar system, there is reason to believe they may be in the asteroid belt where building materials are most abundant. (Carlotto, www.psrw.com) Interestingly, one of the possible ways to detect the presence of an ETI probe, Barrow & Tipler state, is to search for evidence of the waste heat from its construction activities. Such waste heat, it is surmised, would give rise to an infrared excess in the area of the asteroid belt where building materials are most abundant. They seem amused by the fact that much of the observed infrared radiation does, in fact, come from the vicinity of Earth's asteroid belt.

Their interest in this amusing factoid stops far short of labeling it as *evidence of ETI*, however, even though they affirm that such colonization behavior would be typical of any intelligent species with the means to do so. By admitting that probes could be used to observe a society *without its awareness*, the Space Travel Argument undermines its own *absence of evidence = evidence of absence* premise. What argument can be proffered to prove that Earth's biological forms, as well as its biosphere, are *not* the result of the engineering genius of advanced space faring ETI?

Interestingly, the Space Travel Argument suggests that a human being is essentially "a universal constructor specialized to perform on the surface of the Earth." As a matter of fact, all biological life forms on the planet are "universal constructors" from the point of view that they are self-replicating. If the job of a universal constructor is to construct other universal constructors *ad infinitum*, then the appearance of biological life on Earth could be in itself a potential sign of the existence of a extraterrestrial universal constructor. Thus, the Space Travel Argument's circular structure, i.e. *they aren't here because they aren't here*, can be countered with the argument *they are here because <u>we</u> are here*.

Can there be any proof that civilization on planet Earth is <u>not</u> the result of technology which either lost control by the species which planted it or was allowed to, as Barrow & Tipler put it, go *into business for itself*? If scientists believe it possible that an alien species could covertly study any society using a computerized self-replicating universal constructor, how can they be so sure that such an event has not occurred in the past? If Earth were, in fact, an ETI colony, a "preserve" for wild creatures of the Universe, or an environmentally controlled DNA repository, would the creatures in the preserve be any the wiser concerning their origins? Perhaps, then, ETI custodians, or "robot game wardens," are already present in our Solar System.

The next chapter illustrates that the theoretical "Robot Game Warden" hypothesis and the "Zoo Hypothesis" contained within the Space Travel Argument *Against* the Existence of ETI actually have a possible connection to reality.

Chapter Two

Robot Game Wardens:

Evidence of Space Traveling
ETI in the Solar System

With probes it would be possible to study an alien species without it becoming aware of the species which is studying it.

Anthropic Cosmological Principle
Barrow & Tipler

The Space Travel Argument discussed in the preceding chapter concludes that the Zoo Hypothesis is unlikely. If it were true, they state, "our entire Solar System would be analogous to an American national forest, or an African game preserve" and von Neumann probes would be presently acting in the capacity of "Game Wardens." If our Solar System were such a preserve, they surmise, "then all contact must have been rigorously prevented for as long as the robot game wardens were present in the Solar System, since there is not one jot of evidence for any contact in the past." Such a total prevention of contact is, they assert, quite impossible. Surely someone would try to get through; some ETI group would believe contact to be in the interest of Earthlings.

No policing system can prevent all crime, Barrow & Tipler submit, adding that Robot Game Wardens could not possibly prevent a group of ETI from beaming a radio signal to the Earth. (Here they go with the radio signals again!) Furthermore, if an anomalous radio signal was ascertained, how do we know they would tell us about it? The Space Travel Argument presumes to know the language that an otherworld intelligence would use to communicate their presence to us. Many scientists believe that universal language would be mathematics, specifically Euclidean geometry. Thus, it is quite valid to query whether we might be ignoring potential cosmic messages of all imaginable sorts while we await word via radio frequency. For instance, could the huge geometrical fractal-based designs in cereal crops all over the world be cosmic post-it notes from a friendly civilization wishing to make their presence known? The complicated mathematical nature of crop circles has been convincingly demonstrated by researchers.

Ironically, the theoretical "Robot Game Warden" hypothesis and the "Zoo Hypothesis" contained within the Space Travel Argument Against the Existence of ETI have a possible connection to reality. It seems that pictures of such theoretical Robot Game Wardens may have actually arrived via Russian space probe.

The Hollow Moons of Mars?

Compelling evidence of the existence of ETI in our own solar system comes from the strange disappearance of the unmanned Russian satellite probe Phobos 2. In July of 1988, Russia launched two probes, which it

A "sheparding satellite" of Saturn. According to NASA, the shepard is a highly reflective, bright surfaced object. NASA Photo Archives.

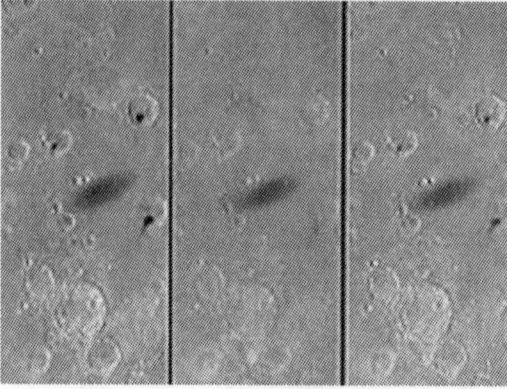

The Global Surveyor caught this Martian landscape image in August, 1999. A NASA press release described the image as the shadow cast by the Martian moon Phobos, and described Phobos as a potato-shaped moon about 8 miles in length. NASA Photo Archives.

called Phobos 1 and Phobos 2, in the direction of Mars with the intention of investigating Phobos, the mysterious, perhaps hollow, moon of Mars. Could this "moon" itself be an enormous bio-dome construction? This question has apparently been asked before, by intelligent people from Earth. One of those people, Dr. Carl Sagan, once calculated from the estimated density of the Martian atmosphere and the odd accelerations of Phobos that this moon is hollow.

Phobos and Deimos, the two moons of Mars, may not be natural satellites, but manufactured ones, writes Bruce Rux in *Architects of the Underworld*. This theory was seriously proposed by Russian scientist I.S. Shklovskii in a 1966 book entitled *Intelligent Life in the Universe*, which he co-authored with Carl Sagan. The authors proposed that Phobos and Deimos should long ago have either crashed into the surface of Mars or careened out of the orbit of Mars due to their small size. The authors also proposed that the center of Phobos was either hollow or was made of ice. It is generally agreed that Phobos behaves contrary to any other moon in the solar system. Phobos is inexplicably gaining speed in its orbit around Mars. Its orbit is circular as opposed to elliptical and it spins unnaturally fast, at about three and a half orbits in one Martian day. In addition, Phobos rises in the west and sets in the east, contrary to most bodies in the solar system.

There is also evidence that some type of activity on the surface of Phobos has been ongoing. For instance, Rux writes, "grooves and track marks mapped on Phobos by Mariner 9 were discovered to have increased in number by the time Phobos 2 arrived at Mars eighteen years later." All of this activity appears to lead to a huge perfectly circular crater on one end which takes up a third of the diameter of the moon. Last, but not least, these two moons were not discovered until 1877, even though astronomers had been observing Mars all along and had sufficient lenses to have observed the satellites if they were there. Some have also wondered whether the deep "grooves" in the surface of this body were caused by cables which tore into it during the transport process. Were these enigmatic orbs placed in orbit by an advanced ET race?

French researcher Jean Sendy wrote a book in 1969 entitled *Those Gods Who Made Heaven and Earth*, positing that, approximately 23,500 years ago, space travelers called the "Elohim" arrived in our solar system in a large hollow sphere. The spacecraft is now called Phobos, and is one of the Martian moons. This author also proffered the theory that on the Biblical "first day," these ETs "stabilized the rotation of Earth's moon so that one side always faced earthward." He wrote

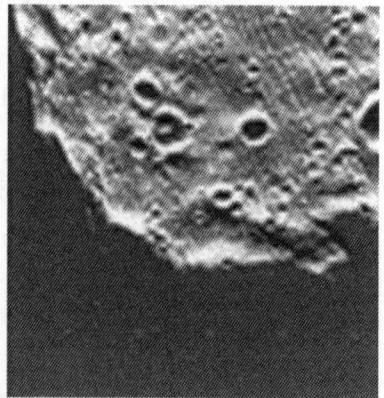

One of the Martian moons, Phobos, behaves differently than any other moon in the solar system. NASA Photo Archives.

that cloud cover which constantly blanketed Earth was removed so that Earth could be closely watched. Eden then became "a climate controlled laboratory," stocked with representative samplings of various life forms. Eden provided "optimum conditions for the development of a superior strain of humans."

Jacques Vallee has admitted that it is *conceivable* that Phobos and Deimos are interplanetary craft which use Mars as a base in their exploration of this solar system. As quoted in *Architects of the Underworld*, Vallee suggests:

> In our present state of the ignorance of the nature of the Martian satellites it is not impossible to think that they are large interstellar vehicles, placed into orbit more than a century ago ... by an advanced scientific community coming from elsewhere in the Universe.

Robot Game Wardens

Let's look at the descriptions that have come in via satellite of our friendly Robot Game Wardens parked just outside of Mars. The goal of the Russian Phobos probe was to fly "in tandem" with the Martian moonlet and drop study instruments onto its surface. The Phobos 1 probe was lost just two months after takeoff due to a "radio command error." The Phobos 2 probe arrived safely in orbit around Mars and began taking pictures. The Phobos 2 probe aligned itself with the moon Phobos, and then suddenly lost contact with Earth. The Russians later released a taped television transmission which the probe relayed in the last few seconds before it lost contact. The transmission was televised in Europe and Canada, accompanied by commentary. The images were described in detail in the April 1989 issue of *New Scientist*. They explain: "the features are either on the Martian surface or the lower atmosphere. The features are between 20 and 25 kilometers wide and do not resemble any known geological formation. They are spindle-shaped and proving to be intriguing and puzzling."

Another photo taken by Phobos 2 and shown on Canadian television shows an infrared scan of a clearly defined latticework of straight rows resembling an underground city. According to John Becklake of London Science Museum, this 60-kilometer city-like pattern "could easily be mistaken for an aerial view of Los Angeles." In addition to these puzzling features on or near the surface of Mars, just before its demise Phobos 2 caught a picture of a clearly defined dark shadow of a thin elliptical object above the surface of Mars. This shape stood out sharply against the surface, and was clearly situated *between* the spacecraft and the Martian surface. As this picture was halfway through, the Russians then "saw something that should not be there."

The Russians refused to release this picture to the public for quite some time until, in 1991, Russian astronaut Marina Popovich smuggled a photo to the press. The last transmission from Phobos 2 was clearly a photo of an enormous cigar-shaped alien mothership, approximately 20 kilometers long, parked next to the Martian moon. It appears that Phobos 2 was knocked out of the picture because somebody didn't want it there. According to Ms. Popovich, the Russians believe it was destroyed, quite possibly with an energy pulse beam.

With respect to these anthropomorphic anomalies, we must also keep in mind that at least two space probes have been inexplicably lost in the vicinity of Mars in the last decade. In August of 1993, just three days before its expected placement in orbit around Mars, the Mars Observer, the culmination of eight long years of work, "mysteriously disappeared." According to *Newsweek's* July 21, 1997 edition, the $980 million probe was "stuffed with scientific equipment."

The next lost-in-space probe, the Mars Climate Orbiter, was officially declared lost on September 24, 1999. *Newsweek* announced that the Orbiter "passed behind the planet, out of radio contact, and never emerged." NASAofficials gave up the search for the $125 million probe, suggesting that it may have entered the Martian atmosphere as close as 35 miles (or 57 kilometers) from the surface, rather than the target altitude of 90 miles (140 k). At that close altitude, the probe would have burnt up in the Martian atmosphere.

However, the same NASA press release suggested it was possible that the craft "had somehow not entered orbit and sped right past Mars on an escape trajectory." Therefore, officials were not certain on 9/24/99 that the probe actually entered the Martian atmosphere, and the story that the craft burnt up in the Martian atmosphere was pure speculation.

Six days later, on September 30, 1999, officials at NASA's Jet Propulsion Laboratory announced that the probe was lost due to metric miscalculations! According to *Astronomy Now*, the Mars Climate Orbiter spacecraft team at Lockheed Martin in Colorado and the navigation team located at JPL in California were using different measurement systems. A peer review board investigating the loss of the craft discovered that one team was using Imperial or English measurements, while the other team was using metric measurements. They claimed that these "human errors" caused miscalculations in the maneuvers critical to correct placement of the probe in Mars orbit. The *Denver Post* reported that Lockheed Martin had been using Imperial measurement and JPL had been using the metric system.

While this explanation may seem plausible, and even laughable, it is remarkable that such an incredibly stupid blunder could be made by a private industry which has for several decades been in charge of multi-million and even billion dollar pieces of equipment. One would think that only a state-run agency would be capable of such incompetent oversight. How many silly excuses can NASA possibly come up with for the loss of countless space probes? And, how lame can these excuses get? Since they know the public needs some sort of explanation, is JPL taking the "human error" fall for an actually unexplainable loss? Could the first explanation have been the correct one? Did the probe not even enter Mars orbit? Was it shot down by hostile forces? We must remain open to this possibility.

Coming on the heels of this speculation was an anonymous e-mail received by Richard Hoagland ("The Real Story," enterprisemission.com) from a person who had an intimate discussion with a Lockheed Martin employee a few days following the loss of the Mars Orbiter. This employee, with top security clearance, was in "complete despair" over the events which led to the loss of the Orbiter on September 24, 1999. He explained that everything had been going along perfectly, and even the most complicated maneuvers had gone "without a hitch." The most important thing, this e-mailer noted, was that the Lockheed employee explained the disconcerting events "using metric measurements in his description." Yet, official press releases repeated by several news sources a few days later, on September 30, declared that Lockheed had been using Imperial measurements in their calculations, and that JPL had been using metric!

Following the loss of the Climate Orbiter, on December 6, 1999 officials at JPL gave up on locating any signals from the lost $165 million Mars Polar Lander. The lost-in-space saga continues...

Artificial Objects on Planetary Surfaces

It would appear that Barrow & Tipler are happy to let any "evidence" other than radio signals stand as evidence of the absence of ETI. Therefore, they are awaiting "word" in a language understandable to materialist scientists, while ignoring any other symbols, signals or sigils which might arrive via an alternative route. Cosmic signals from ETI might just as well be evidenced by physical structures we have chosen to explain in other ways, such as the complex mathematical fractals which make up crop circle designs, Old and New World pyramid structures which appear to be built as aeronautical guides, and which happen to be geometrically-aligned with the complex architecture of the Cydonia region on Mars. (see Hoagland, Carlotto, McDaniel, et al.)

A paper written by Richard Hoagland and Erol Torun entitled "The Message of Cydonia: First Communication from an Extraterrestrial Civilization?," (www.enterprisemission) discusses the implications of the humanoid face and other anthropomorphic objects, including several pyramids, which have been noted on Mars. Hoagland suggests that the presence of these objects is trying to "tell us something." The discovery of a redundant "tetrahedral geometry" which is "encoded" at this Martian site has led this body of researchers to conclude that "every major energy center—on the sun, and on most of the planets and their active satellites—emerges at the surface in conformance with the predictions of an embedded tetrahedral model, primarily at either 19.5 degrees north or south latitude."

These researchers see this embedded structural geometry as a "deliberate, technical communication." They write: "common sense dictates that, whatever the origin and culture of the 'senders,'their purpose clearly would have been to communicate something of fundamental significance." They suggest that the "Cydonia Message" may be a "unified field" message, aimed at showing us a new energy resource which would result in "a near-term, dramatic breakthrough in fundamental space propulsion technology, with obvious implications for space exploration and eventual settlement." The possible breakthrough applications of an "unexplained but demonstrable physical phenomenon in operation throughout the solar system," they believe, is an urgent reason for immediate verification of the Cydonia message.

A paper published in 1996 by Mark Carlotto and Michael Stein entitled "A Method for Searching for Artificial Objects on Planetary Surfaces" (*www.psrw.com*) concludes that the existence of ETI can certainly be ascertained in ways other than tuning in to radio frequencies. They write:

> The focus of the search for extraterrestrial intelligence (SETI) has been to look outside our solar system at radio frequencies for signs of intelligent life. Such a strategy is consistent with current information suggesting that it is unlikely that intelligent life could have evolved on the other planets in our solar system. Our knowledge to date cannot, however, rule out the possibility that extraterrestrials or their probes may have reached this solar system. If so, they may have altered planetary surfaces in ways that are detectable through remote sensing.

In this paper, which describes anomalous geometrically-aligned structures on Mars including the now infamous "Face," Carlotto and Stein outline a strategy to detect non-natural objects on planetary surfaces based on the fractal modeling of terrain. Due to the self-similarity properties of fractals (i.e. they look the same across a range of scales or resolutions) areas of terrain that are dissimilar could be detected. Thus, by comparison, such anomalous or "man-made" objects would effectively jump out of the terrain.

The problem is, NASAdoesn't appear interested in showing the public what it finds. The circular nature of the Space Travel Argument denies the validity of any analysis of such artificial objects on other planetary bodies with its *a priori* stance that there is not *one jot of evidence for any contact in the past*. Could this be due to the warning put forth by the Brookings Institute against public disclosure of any discovery in the solar system which might be construed as an extraterrestrial artifact?

The *Brookings Report*, officially entitled "Proposed Studies on the Implications of Peaceful Space Activities for Human Affairs," was commissioned by NASA in 1959. The report by the infamous Think Tank, dated April 18, 1961, was presented to the First Session of the 87th Congress. It warned that artifacts left by "life forms" on planets investigated by Earth's unmanned space probes would lead to the "disintegration of civilization." The document recommended complete censorship of any future discovery of extraterrestrial artifacts, based on anthropological studies of societies which have disintegrated in the presence of culture shock. It specifically feared that "religious fanaticism" would wreak havoc on social institutions. This exaggerated response has most likely been the source of the seemingly premeditated obstruction of truth with regard to the discovery of possible extraterrestrial artifacts in our solar system.

The Man in the Moon

Even closer to home, circumlunar anomalies have been noted with regard to Earth's moon over the years. The first oddity is that we don't have a name for the object that seems to "peer" down upon us at night. Instead we use the generic term, the Moon, which implies ownership. It's "ours," it circles around our Earth, we attribute cause and effect occurrences to its peculiar emotional sway. The Moon belongs to the Earth as some kind of package deal. But what if we discovered that it *isn't* "ours," but actually belongs to someone else? Would we turn around and hi-tail it home once we found out? In fact, this appears to be NASA's behavior since the last Apollo mission in 1972. What scared NASA? Was it the Man in the Moon?

The Moon functions as a convenient brake on the Earth's rotation, pulling on its oceans and slowing it down. The Moon's gravitational pull on the Earth stabilizes the Earth's wobble and regulates it's climate. The smallest change in the Earth's tilt could cause serious climatic turmoil. In this, the Earth depends upon the Moon for its ability to support life. The Moon's rotation around the Earth is perfectly synchronous with the Earth's rotation. The Moon turns once on its axis in the same time it takes to revolve once around the Earth. The peculiarity in the axis tilts of both the Earth and the Moon causes one side of the Moon to remain hidden from our view at all times. Nothing says nature cannot work like "clockwork," but the reason for this synchronicity has eluded scientists. The Moon is one-quarter the diameter of Earth, but has only one-eightieth of the Earth's mass. It has been noted that the Moon "lacks the proper composition for it to have assumed such an orbit."

According to an earlier analysis of the Moon's motion by NASA scientist, Dr. Gordon McDonald, the Moon is hollow. However, he could not accept that his figures were correct because he could not accept the theory of a hollow Moon. Like it or not, other scientists reached similar conclusions. In the February, 1962 issue of *Astronautics*, it was reported that the Lunar Orbiter experiments had discovered "the frightening possibility that the Moon might be hollow." In his 1966 book, *Intelligent Life in the Universe*, Carl Sagan stated that, "a natural satellite cannot be a hollow object." In agreement with this basic scientific premise, Russian scientists, Vasin and Shcherbakov, published an article in *Sputnik* magazine which stated that: "the Moon is not a natural satellite of Earth, but a huge, hollowed-out planetoid fashioned by some highly advanced, technologically sophisticated civilization."

In his 1975 book, *Our Mysterious Spaceship Moon*, Don Wilson described the following peculiar incident. In November of 1969, at lift-off from the lunar surface, the Apollo 13 rocket booster crashed back to the surface, striking with the force of 11 tons of TNT, approximately 87 miles from the Apollo 12 seismic equipment which had been left on the surface. The Moon vibrated for "more than three hours at a depth of up to 25 miles." Wilson suggested that "these gonglike vibrations puzzled NASA scientists, but are explainable if the Soviet theory of an inner metallic spaceship hull is correct."

The issue of the "hollow moon" was finally put to rest on March 16, 1999, when NASA scientists issued press release 99-43 disclosing that lunar data support the idea that a cosmic collision split the Earth and the body which later became its satellite. The press release stated: "analysis of data from NASA's Lunar Prospector spacecraft has confirmed that the Moon has a small core, supporting the theory that the bulk of the Moon was ripped away from the early Earth when an object the size of Mars collided with the Earth." NASA scientists, who presented these findings at the 30th Lunar and Planetary Science Conference in Houston, indicated that the lunar core contains less than four percent of the Moon's total mass, and may even be less than two percent of its total mass. In comparison, the Earth's iron core constitutes about 30 percent of the planet's mass.

Papers presented at this conference also indicated that "similarities in the mineral composition of the Earth and Moon indicate that they share a common origin." However, it was noted that, if they had simply formed from the same cloud of rocks and dust, the Moon would have a core similar in proportion to that of the Earth's. Based on information obtained during the Apollo era, the press release stated, it was suggested that a "Mars-sized body" hit the Earth in its earliest history *after* its iron core had formed. The impact ejected rocky, "iron-poor material" from the outer shell into orbit, which collected to form the Moon, and was then caught in orbit around the Earth. It has also been discovered that "a broad section of the southern far-side of the Moon has large localized magnetic fields in its crust," indicating strong magnetized concentrations on one side of the Moon. This would explain why the Moon always has one side exposed to Earth.

Other circumlunar anomalies have been reported since the invention of the telescope. In his book, *Penetration*, Ingo Swann points out that, except for the state-of-the-art telescopes in "official" hands, most telescopes owned by private individuals are of fairly low-resolution. However, fairly obvious "luminous phenomena" have been noted on the Moon by thousands of individuals worldwide. But, since these individuals are marginal to officialdom, their evidence is laughed out of town. Even NASA published a chronology of some of these events in 1968 (NASA Technical Report R-277). This catalog documents 579 lunar oddities which were noted between 1540 and 1967, 75% of which are characterized as "luminous phenomena." Many of these lights have been seen moving, and forming geometric patterns. The crater Plato is the most famous for lights, having a disco dance floor sixty miles across which is lit up and even changes color. In addition, beams of light have been observed, as well as glowing objects which move around on the surface.

According to Swann, in addition to moving lights and lit-up craters, huge dark objects have been seen moving "slowly" across the face of the Moon. One of these sightings was described on October 12, 1954 by an astronomer at the Edinburgh Observatory. He described it as a "dark sphere" which traveled in a straight line from one large crater to another. It took 20 minutes for the sphere to travel between the two craters, both of which were known for exhibiting light phenomena. It has been calculated that the object was traveling approximately 6,000 miles per hour and had to be nearly four or five miles in dimension. A similar object was seen in September of the same year. This time, the large sphere left the lunar surface and moved out into space.

In 1950, a "mechanical object" was observed on the Moon by British astronomer H.P. Wilkins, which he described as "some type of glowing machine hovering near the crater floor." In his book *Our Mysterious Spaceship Moon*, Don Wilson uncovered 400 such reports of odd lunar observations. He wondered: "is the Moon inhabited as a base used by extraterrestrials to visit Earth?" Other lunar abnormalities noted by Wilson were the detection of radio signals in the vicinity of the Moon circa 1927, 1928, and 1934. Also, in 1956, "code-like radio chatter" was reported to have come in from the vicinity of the Moon. In addition, in 1969, the Apollo 11 astronauts reportedly heard strange "radio noise," described as "loud sirens and buzz-saw sounds," as they approached the lunar surface. Upon hearing these noises, Mission Control reportedly asked "you sure you don't have anybody else up there with you?" Apollo 12 astronauts reported the sighting of two "bogeys," NASA's terminology for UFOs, as they neared the Moon. While on the Moon's surface, Apollo 15 astronauts witnessed various white objects which appeared to have been propelled or ejected. These have been shown widely on television. (see *Cosmic Test Tube*)

There have been many reports of quite large "anomalous" objects moving across the lunar surface in a pattern which implies "directed flight." Smaller "glowing" objects have also been seen flying in and out of large craters in a type of flight pattern. Do these craft constitute an "occupational hazard" in terms of our occupation of the Moon?

According to Swann, before officials realized the presence of certain unnatural details on the lunar landscape, a few early high-resolution photographs landed in the hands of capable photo analysts. In particular, two important NASA photos somehow eluded the early detection of the airbrushing reality-engineers at NASA. These pictures were published in a 1975 book entitled *Somebody Else is On the Moon,* written by George Leonard. These two photos are described by Ingo Swann in his book *Penetration.* In July, 1969, Apollo 11 caught a clear picture of a "glowing, cigar-shaped object" close to the lunar surface, which left a "vapor trail." (NASA photo No. 11-37-5438). In July, 1972, Apollo 16 caught another picture of a large cigar-shaped object, glowing white, and casting an elongated shadow on the surface of the Moon. (NASAphoto No. 16-19238).

Not since these two pictures has NASAmade any more such blunders. High resolution telescopes have been in place since 1959, but no high resolution lunar visual information has been made public. In addition, NASAhas not released to the public the Orbiter photos of certain lunar areas known to be particularly active with light phenomena. The Pentagon probe *Clementine*, under the auspices of the Ballistic Missile Defense Organization (Star Wars), which was equipped with infrared, ultraviolet and other types of high-resolution cameras for lunar mapping, was inexplicably lost. About a tenth of its pictures were withheld from the public. As a matter of fact, writes Swann, NASA has accumulated approximately 140,000 high-resolution lunar photos, none of which have been shown to the public. All the public is ever shown, Swann rightfully complains, are "low-resolution" photos, retouched or cropped where necessary. NASA has not released anything with the clarity we should be capable of obtaining with "Spy in the Sky" technology.

In support of this allegation, an independent comparison of officially released images on the web from *Clementine* suggests that some of the images have been altered. (See articles at www.lunaranomalies.com). A commentary entitled "Evidence for Faked Clementine Images," indicates that the completely featureless, smooth (i.e. allegedly airbrushed) image of the crater Plato, one of the most significant and possibly active lunar features, suggests to amateur astronomers "that there is something in the crater Plato that someone doesn't want us to see." Attempts to gain other pictures of this crater came up negative. There is a "gap" in the available images around the center of the crater.

Another independent study, entitled "Enigmatic Lunar Structures," was undertaken by Steve Troy. He indicates that there are some natural features on the Moon that simply look unnatural. But, he asserts, other structures, as Hoagland has suggested, take the form of "crystalline rebar and support structures that extend above the surface." These structures appear to be remnants of structures which were once much larger and have deteriorated due to meteor impact. Studies have shown that glass could be used as building material on the Moon, and in a lunar vacuum it would have the structural strength of steel. Troy concluded from his painstaking studies of Apollo catalog photos that these hard lunar geometric surface structures appear to be "architectural." He notes that the refractive filamentary structures contain highly textured patterns, "redundantly repetitious right angles," as well as "triangular definition" and even

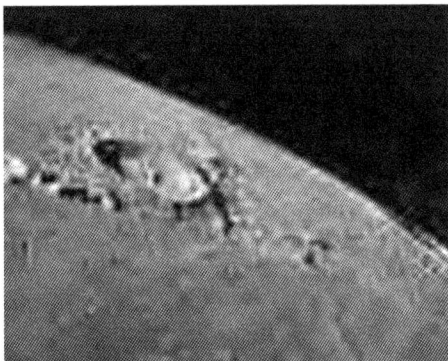

The crater Copernicus and what appears to be an anomalous "box-like" dome structure within. NASA Photo Archives.

"arches." The structures are raised above the lunar surface. Such architectural and foundation-type designs are found in and around craters. He indicates that some anomalous structures which are found out of the path of deposition and wasting are found to have walls and buttresses higher than the natural terrain. The author concludes, "there are areas on the Moon that simply cannot be explained geologically." He explains, much of what he has studied in lunar photos "should not be there according to the traditional model of natural stratigraphy and geology.

Another independent study of lunar features called "The Case for a Dome Near Copernicus," discloses that a "genuine artifact image" shows a box-like dome above the crater Copernicus. The architecture of this "dome" is consistent with Hoagland's "Sinus Medii Dome." In front of this newly discovered crescent-shaped dome is something semi-translucent, which makes the dome appear as if it's underwater. This structure might be some type of ancient biosphere. In addition, an independent panel called the Lunar Artifacts Research Group has confirmed the presence of all of the anomalies described in detail by Hoagland, including the Shard, the Tower, and the Castle.

This is a curious and disconcerting state of affairs. What is NASA trying to hide? According to George Leonard, a highly advanced underground civilization has been actively engaged in a full-scale mining and manufacturing Colony on the Moon. And I don't think their name is O'Neill.

Bucky Fuller on the Moon?

On November 22, 1966, *The Washington Post* ran a front-page headline which read: "Six Mysterious Statuesque Shadows Photographed on the Moon by Orbiter." The *Los Angeles Times* also picked up the story a day later. In his book *The Gods of Eden*, William Bramley discusses the *Post* article, which described a photo taken by the Orbiter as it passed over the surface of the lunar landscape from a distance of about 20 to 30 miles. The *Post* described the photo as revealing the shadows of six spires, arranged in a purposeful geometrical pattern, situated inside an area called The Sea of Tranquility. The shadows seemed to indicate that the objects were cone-shaped or pyramid-shaped. It was even noted by a Russian scientist in *Argosy* magazine that these shapes resembled the great pyramids of Earth. The Viking mission in 1976 again showed these pointed objects.

Richard Hoagland of the "Mars Mission" believes that the first Apollo mission made the discovery that we were not the first to step foot on the Moon, and this discovery immediately halted further manned exploration. He offers some intriguing observations regarding the strange Soviet and American behavior of the late sixties. According to the Summer 1995 issue of Hoagland's *Martian Horizons*, inexplicable Soviet behavior began in the late 1960s simultaneously with the beginning of NASA's strange behavior with regard to the Moon. Hoagland writes: "This recently revealed, paradoxical Soviet behavior, so at variance with all their Cold War rhetoric, includes a major mystifying incident, which occurred at the height of the so-called space race." This peculiar incident was the scheduling and abrupt cancellation of the first planned Russian flight of cosmonauts around the Moon in December of 1968, which would have occurred just days before the infamous Apollo 8 journey. This crucial Soviet cancellation, which has since been ascertained from KGB documents, was ordered "on the eve of an imminent 'win' in the East/West 'Moon race,' [and] came from the very top of the Soviet leadership." Hoagland believes that the timing and illogical order of this behavior is bizarre and inexplicable.

Hoagland further observes that "this dramatic Soviet decision occurred immediately after a major unmanned Soviet mission to the Moon ... [which] returned its high-resolution films directly to the Earth." These high-resolution films were matched later by similar images which were returned to the NASA photo lab by the Apollo astronauts. Hoagland wonders if the Russian lunar analysts discovered something on their high-resolution lunar films ... "something that caused the sudden, apparent paralysis history now records regarding all further Soviet policy *vis a vis* the Moon." This sudden paralysis was later inexplicably repeated in the American space program. Hoagland wonders what went so radically wrong with these space programs and what has kept "not only the Americans and Russians, but even the Europeans, Chinese, Indians, Brazilians and Japanese so closely 'by the shore'." Could it be, he wonders, that we simply learned that we were not the first intelligent beings to reach the Moon? He asks, has some intelligent race already left "a dizzying array of dazzling, monumental artifacts on satellites and planets all across this solar system?" Furthermore, did the conclusions of the *Brookings Report* admonish "a hasty retreat back to our own planet," forcing us to "stay home for the next 30 years, frightened out of our wits?"

In his video *The Mars/Moon Connection*, Hoagland seems to affirmatively answer these questions. He provides an overview of the developments which comprise his team's discovery of man-made structures on the Moon and Mars, and their mathematical alignments with the pyramids of Earth. Hoagland asserts that between 1966 and 1972, Lunar Orbiter, Surveyor (the first robot explorer) and Apollo missions found and corroborated "the real Moon," which perhaps only a few people at the head of NASA have ever seen. Hoagland's Moon story begins with the Surveyor 4 mission, which was cut off abruptly with no indication of trouble. Hoagland explains, in one microsecond Surveyor 4 was there, and in the next it was gone. The engineers ruled out an explosion, because there was no evidence in their instrumentation of an explosion. Hoagland explains that the engineers would have heard the echoes of the death throes on the telemetry link if there had been an explosion. NASA has insisted only that "the mission failed," and that it was "lost during a retro rocket burn." NASAappears to be very sensitive about this issue. Hoagland's explanation for this abrupt loss of signal is a 6,000 mile-an-hour crash into the walls of an ancient high rise Crystal City on the Moon. And he even has pictures to prove it.

The Apollo 8 space mission made ten orbits around the Moon, and returned hundreds of frames which were not shown to the public. Apollo Frame 4856 shows double craters, or pairs of holes, aligned in geometric rows like a "trusswork" of steel girders. Over this support structure appears to be the remains of a Bucky Fuller-style geodesic dome which may once have been a complex biosphere which protected occupants of a previous colony from the hostile, meteor impacted surface of the Moon. There are vertical and horizontal ruins, showing parallel patterns which are mindful of a 3-D cross section of the ruins of Hiroshima. Some of the structures appear to be about 20 to 30 feet tall, and have edges which refract sunlight.

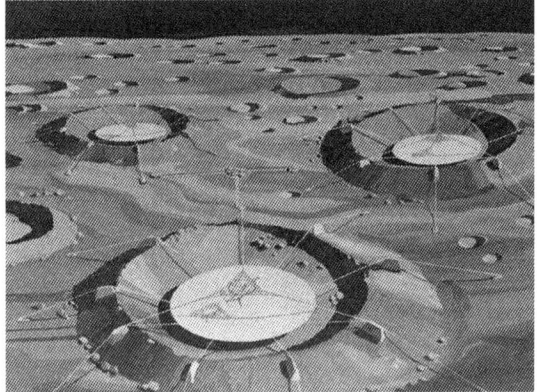

The Apollo 10 picture catalog contains several pictures that are simply black. Hoagland decided to order one of these pictures. The frame which came back, Frame 4822, shows a grid-like pattern resembling the ancient ruins of a city. This

Artistic design of moon colony utilizing array of SETI antennae on the Moon. Artist: Rick Guidice, 1976. NASA Photo Archives.

frame also exhibits the light refractions of a structure resembling what Hoagland calls a 3-D Crystal Palace suspended thirty miles above the ground, which appears to be about ten to fifteen miles long. Hoagland surmises that Surveyor 4 slammed directly into this structure.

Hoagland has received many pictures over the years from sources at NASA and other agencies which he and his team are in the process of analyzing. The structures he has studied in detail contain shapes that are not found in nature: trapezoids, hexagons, conical shapes, spires and dome shapes built into geometric patterns. The edges of some of the structures seem to sparkle as if they are made of a transparent crystalline material which reflects sunlight. Hoagland is incredulous that the astronauts didn't see this enormous matrix of structures, which he maintains is "as big as a mountain." However, as the *Post* reported in 1966, the Lunar Orbiter *did* see something resembling the shadows of six tall spires as it passed over the surface of the lunar landscape.

In 1961, President John F. Kennedy announced that we were going to colonize the Moon in order to have a base from which to explore our solar system. A "space race" began between the two Earth superpowers. The plan to colonize the Moon was abruptly abandoned in about 1972. Apollo 17, in December of 1972, was the last American Moon flight, and Luna 21, in January of 1973, and Luna 24, in August, 1976, were the last Soviet Moon flights. Thereafter, the two superpowers, who just happened to be common enemies, elected to put down their ideological differences to cohabit a deluxe apartment in the sky. What's wrong with this sitcom?

An artistic study of a habitable lunar base by Rick Guidice, 1978. NASA Photo Archives.

The official reasons for scrapping the Apollo program were reiterated on a television report in May, 1999. The reasons were listed as "lack of purpose, safety fears, and budget cuts." What could they mean by "lack of purpose"? Exactly what do these "safety fears" allude to? Isn't it just as dangerous to build a platform in the sky? It was also reported on this television program that the international 16-nation space station currently under construction will end

An artistic study of a Zero-Gravity biodome habitat by Rick Guidice, 1978. NASA Photo Archives.

up costing in excess of $100 billion.

If Apollo was cancelled due to "budget cuts," where are we getting this money? It was then reported in May, 1999 in the *Washington Post* that the Russian space station *Mir* is to be abruptly abandoned by the end of 1999. The article made it clear that the US was hopeful that Russia would abandon the failing space station in order to contribute funds to the new space station, which has yet to be named at this writing. What is the urgency behind this unprecedented international endeavor? Obviously, the Russians do not have the stable economy necessary to contribute to this extraordinary enterprise. Does this situation essentially comprise a global necessity of serious magnitude? In light of all the "Armageddon" type films of late, about asteroids and comets hitting the Earth, is there something coming our way?

The International Space Station is being erected at enormous cost to the public, yet we were never told why we were unable to use the "natural" satellite we already have. As Hoagland and others have pointed out, the circumstances of this historical pattern are certainly peculiar. It would appear that the pattern of information disseminated by Earth's hierarchy to its lower echelon peons consists of nothing but circumlunar circumlocution.

An artistic study of an underground habitat on Mars by Garrett Moore, 1991. NASA Photo Archives.

The Project Horizon Moon Base

In his book, *The Day After Roswell*, Col. Philip Corso writes that U.S. Army plans for a Moon base began in about 1959 and were set for completion in 1967. The reasons for the proposed Moon base were (1) to establish a presence on the Moon before the Soviets, (2) to set up early warning surveillance against Soviet missile attack, and (3) surveillance and defense against alien space craft. Corso writes that Project Horizon was a military plan to establish "a skirmish line in space," in order to protect the Earth against a surprise attack from incoming hostile alien ships. This military project was eventually sidetracked when space exploration was handed over to the civilian-based NASA.

In 1961, NASA agreed to cooperate with the military in covering up the truth of the alien threat. They agreed to open up a "back channel" of communication on board space craft explicitly for reports of any hostile activities conducted by the EBEs, including shadowing or surveillance activities. In 1960, Corso's book maintains, military surveillance satellites were covertly placed in orbit in order to pick up evidence of alien activities in remote regions of the Earth and to keep the Earth under constant surveillance in case the ETs decided to set up terrestrial shop. He asserts that Cold War strategy was used as a cover to keep an eye on "other" hostile encroachments. Since we didn't want to let the Soviets know we were utilizing Earth orbit for spy satellites, the trick was to get a satellite up there in secrecy. Thus, the Lockheed-designed "Corona" spy satellite was covertly loaded into the NASADiscoverer rocket without the American press finding out.

Project Horizon was to begin with a crew of fifteen workers, who were supposed to set up a pre-fab sub-lunar outpost, which was to include a communication system relying on geosynchronous Earth satellites to relay transmissions to Earth. One of the main purposes of the lunar base was to maintain constant vigilance of the Earth and to track any vehicles spotted in space. According to Corso, the planned launching pad was to be located in a secret area in Brazil. It was also planned that the lunar base would be backed up by an orbiting space station, as another component in what he asserts was an "elaborate defense against ETs." The space station would also provide a safer platform for particle-beam weapons testing. Corso maintains that Project Horizon became the ultimate model for the construction of both the Russian *Mir* space station and the American *Freedom* space station, since it taught us how to plan construction projects in a weightless and airless environment.

Corso writes that the "Cold War" provided the budget to defend the Earth against hostile invasion by ETs. But what made them sure that the alien presence was hostile? From the beginning of the space race, he maintains, the ETs were actively interfering with launch vehicles. In some cases, they aggressively "buzzed" manned and unmanned space vehicles, jammed radio transmissions, and caused electrical and mechanical malfunctions. Astronauts were specifically instructed not to report any UFO activities they witnessed during space flights. The transmission signals between space vehicles and NASA were scrambled so that certain communications would not be heard by anyone else. Due to these threatening behaviors, Corso claims, NASA had to "rethink astronaut safety" in the Mercury and Gemini space programs.

The Alaskan HAARP Project

While the Army planned a Moon base, the Navy had its own pet project in mind, explains Corso. The Navy was interested in undersea bases, anti-submarine defense, and tracking of unidentified undersea objects. However, Project Horizon was impeded by NASA, which had its own agenda. Because of the switch to civilian management of space exploration, Corso explains, the U.S. never built a permanent base on the Moon. However, it would appear that the Alaskan HAARP Project, completed in 1995 by the Air Force, is part of this overall plan emanating from the perception of alien hostility. As Canadian journalist, William Thomas, writes in *Paranoia's* Fall, 1995 issue, the military's secret plan to "pluck" the Earth's ionosphere like a "celestial HAARP" will orchestrate "potentially catastrophic events over the heads of North Americans, and other 'enemies' as yet unspecified." The "High-frequency Active Auroral Research Project" is a four-acre antenna system, situated in the Alaskan wilderness, which includes an array of 72-foot towers which will be plugged into a 320,000-watt transmitter.

Thomas writes that the Air Force is planning scores of these "electromagnetic concerts" with various code names. In addition, the Office of Naval Research, a partner in the HAARP enterprise, is interested in the capacity to carry messages to deeply submerged submarines. The HAARP system was devised to be aimed at the sky, but the Department of Defense wants to increase funding in order to aim the transmitter back to Earth. This "Earth tomography mission" will turn HAARP into a very dangerous x-ray machine, capable of "exposing underground tunnel and bunker networks anywhere in the Northern Hemisphere." Thomas wonders who might be the "real owners of these mysterious underground complexes the U.S. Air Force is so eager to find." Unfortunately, while x-raying half the planet to "illuminate such alien notions," HAARP will also repeatedly scan all living organisms with dangerous electromagnetic energies, wreaking serious atmospheric and genetic havoc, greatly accelerating cancer formation and brain tumors, trashing immune systems, and causing birth defects and other cellular mayhem.

Despite warnings of serious planetary perturbations, the military has gone ahead with this highly secretive and infinitely dangerous project. But what is its ultimate purpose? Corso claims the Army and the Air Force had in hand at least 122 photos indicating some evidence of an alien presence on the Moon. Corso writes that the EBEs were colonizing "our" lunar surface, and that they attempted to keep us away from "their" Moon base by buzzing our space vehicles every time they got near. Army intelligence analysts also speculated that "the Apollo Moon landing program was ultimately abandoned because there was no way to protect the astronauts from possible alien threats."

As Philip Corso has outlined in his book, the Project Horizon plan for a Moon base never reached full fruition, but it did teach us one thing: how to plan huge engineering projects in a weightless and airless environment. He also asserts that the lunar base was to be backed up by an orbiting space station, as a secondary component in the defense against hostile encroachment by ET visitors. The military being the military, they assumed the ETs were hostile. Of course, we can now only guess whether this reaction made the situation worse, or saved our sorry asses. Corso chooses the latter. The proposed space station was also to provide a safer platform for potentially dangerous particle-beam and atomic weapons testing. Corso maintains that the Project Horizon engineering plan was the "model" for the construction of both the Russian and American space stations.

Essentially, there is enough information in this quagmire to speculate that the International Space Station is an extension of Project Horizon, which essentially moves the space platform from secondary position to main defense, with Project HAARP as ground defense, against invasion by extraterrestrials as well as freewheeling asteroids and comets.

Let's abandon the pattern of circumterrestrial circumlocution and say what we mean. The Moon may be inhabited by advanced extraterrestrial construction engineers. Is that why we turned around and came home, and began working on our own "little" $100 billion project?

Moon Hoaxes

The idea that "somebody" is on the Moon is interesting in itself, but not so utterly perplexing as the opposing scenarios concerning "who" that somebody might be. The competing theories only serve to muddle the issue, so it is obvious that this confusion has been engineered, probably at the "CIA Weird Desk." (Yes, such a thing does exist.) A quick overview of these competing theories is in order.

Alternative 3 was a British "docudrama" which alerted British citizens to a bizarre scenario encompassing a shared US/Soviet Moon base, and the abduction of American and British scientists as the labor force for this Moon colony. The May, 1999 issue of the Australian magazine *Nexus* contains a letter from Leslie Watkins, producer of this 1977 "docudrama." Watkins admits that this television show was a "hoax," and then goes on to say that since the show elicited such a tremendous public response, he later became convinced that there must have been an element of truth in it. This is total bunk. Since Orson Wells' 1938 "War of the Worlds" radio broadcast, our friendly planetary tricksters have been keeping the public primed for hysteria when it comes to the subject of ETs. There is no reason to equate this primed public response to concepts of "truth."

As a prelude to the Project Horizon Moon base, Corso states, a total of 756,000 pounds of construction materials were *to be* shuttled to the Moon between April 1965 and December 1967. Oddly, the Alternative 3 scenario took into account that some of this was the stuff that the space "shuttles" were shuttling, for, after all, what is the purpose of a shuttle? But the spin on this story was that the Colony on the Moon was a shared US/Russian base on the "dark side." However, an opposing theory is that we never even stepped foot on the Moon. The latter is an intriguing theory.

As Ralph Rene explains in his book, *NASA Mooned America*, it is apparent that several of the most famous NASA photographs of astronauts on the Moon were artistically contrived pictures taken in a studio, and further doctored in government photo labs. Rene describes these contrived photos in detail in his book and in a Winter 1997 interview in *Paranoia* magazine. Rene also doubts that we were capable of manned flights beyond the Van Allen Belt, which serves as a solar radiation shield, since we had not yet developed the necessary light-weight radiation shielding for space suits. Rene suggests nobody is on the Moon at all, and that there are no ETs. He suggests that the reason for the hoaxing of the Apollo Moon landings was to appear to be spending the money allotted to NASA, rather than giving it back and admitting to the world we couldn't land a man on the Moon.

The points Rene makes are convincing, yet they hinge on a few statements made by Col. Corso. As he asserts, it would appear that the development of "supertenacity" fabrics may have come out of the Roswell crash, but Corso does not indicate whether these materials were adequate to serve as radiation shielding. However, it is clear that companies like Monsanto and Dow were working

Tranquility Base on the Moon. Why are there no stars in space? Two competing theories are that we never went to the moon and the pictures are NASA dupes that didn't include stars because the constellations were too difficult to reproduce, or we went in order to investigate ancient biodome structures, which have been airbrushed out of the picture. NASA Photo Archives.

Houston, we've got a problem. Hoagland believes that the reflection in the mask is an ancient artifact or structure, and Rene believes the reflection is NASA's studio equipment. NASA Photo Archives.

on materials which would be useful for the defense industry, and, as Corso suggests, these developments ultimately lead to "stealth" technology. It is not clear, however, that adequate and lightweight radiation shielding would have been developed as early as the 1960s. Therefore, for this and other strong reasons, it is quite possible that Earth humans have never actually stepped foot on the Moon, and that the Moon colonizing activities we have observed via satellite and high resolution telescope are of extraterrestrial origin.

Actually, just how long have military and government higher-ups known about the presence of huge ETI spaceships in Earth's atmosphere? According to Jim Marrs in *Alien Agenda*, while testing new long-range radar equipment in 1953, Air Force operators picked up a huge object circling in orbit about 600 miles above the Earth near the equator. The speed of this enormous "satellite" was clocked at 18,000 m.p.h. A second large object was then detected about 400 miles out, and was tracked on radar. The detection of these two enormous "artificial satellites" touched off several news reports which necessitated a cover story by reality-engineers. While the combined armed forces erected Project "Sky Sweep" at White Sands, NM, the cover story that these were small "moonlets" was simultaneously erected. *Time* magazine published a story called "Second Moon" in March, 1954, explaining that small moonlets, which were similar to asteroids, had entered Earth's orbit. Various stories followed about how tiny moons in orbit around Earth might someday be used as artificial satellites. Then in 1954, NASA stated it had picked up strange signals from an unknown orbiting object. French astronomers also corroborated this report.

The "moonlets" and the "signals" faded into history as our Orwellian reality engineers somehow convinced us that 2+2=5. It is clear that the military has been aware of the ETI presence for quite some time, having been

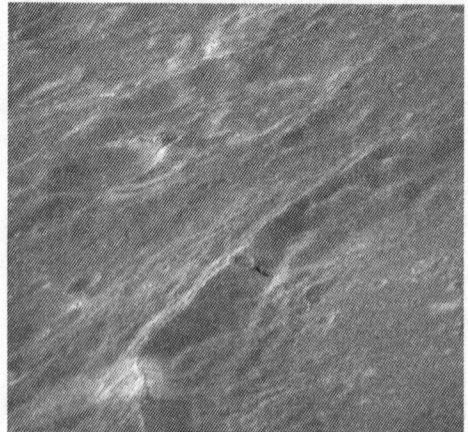

Near the center of this picture is an anomalous object casting a shadow along the edge of the Lobachevsy Crater on the Moon

actively pursued and "buzzed" by ET craft in our own atmosphere, and during early unmanned space flights. Could it be that the reason the astronauts didn't see Hoagland's "enormous matrix of structures" on the Moon is because they are, in Rene's words, *astronots*? Or, could it be that the first Apollo mission put an end to any further manned exploration? In any case, it would seem that it was not necessary to *go* to the Moon in order to see what was going on there. The confusion is whether there was one actual Moon landing and the rest were faked, or whether

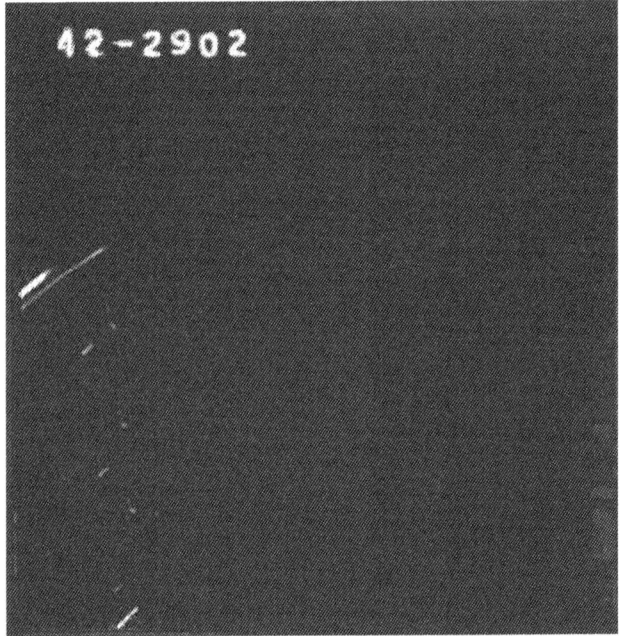

Anomalous light pattern in the atmosphere of the Moon. Johnson Space Center image.

Earthlings never stepped foot on the Moon at all. The truth of the matter rests on whether the technology for lightweight radiation shielding existed at the time, and if it did not, why weren't the "astro-nots" dropping like flies from cancers and leukemia?

Is it possible that it was deemed too dangerous to attempt actual manned Moon landings, and NASA and the military chose to "deep pocket" the Apollo budget funds and fake the whole thing? Philip Corso has made some interesting statements with regard to the relationship between NASA and the military-based "Roswell working group." He explains that this group established and enforced policy at "above top secret" level. The working group's mission, Corso writes, was the "camouflage of our discovery of alien life-forms visiting and probably threatening Earth." The working group believed that any expedition into space, especially for the purpose of establishing a permanent military presence, had a high probability of hostile encounters with ETs. If a military exchange occurred in space, it would be very difficult to keep this news out of the press.

Corso writes that the Roswell working group decided to put space exploration in the hands of a civilian agency, NASA, but kept initial control over its bureaucracy and hand-picked its personnel. This turn of events led to a "byzantine bureaucratic struggle" between members of the same organizations who held different levels of security clearance, as well as different policy objectives, and knowledge of historical precedents. As Corso sees it, underlying all of this turmoil was the assumption that the world was not ready to learn the real truth about the existence of extraterrestrials and, at least according to him, "the likely threat these cultures posed to life on Earth." Corso claims that civilian (NASA) plans called for an outpost in Earth orbit, not a Moon base, and so the Moon base plans were eventually scrapped.

In light of all this, it is possible that these reality spinsters erected the bizarre cover story called Alternative 3, so that any activities noticed on the Moon would be attributed to ourselves. This scenario would serve to cover the fact that we never even set foot on the Moon, and provides a cover story for a huge mis-appropriation of funds which took place under the Apollo cover. It also serves to maintain the necessary illusion that there are no extraterrestrials.

The Earth Without the Moon

According to Immanuel Velikovsky, in his unpublished manuscript entitled *In the Beginning* (www.velikovsky.collision.org), the ancient Greek philoso-phers Democritus and Anaxagoras taught that there was a time when the Earth was without the Moon. Aristotle wrote that the aborigines of Arcadia, the Pelasgians, occupied the land "before there was a Moon in the sky above the Earth." Apollonius of Rhodes also spoke of the early Arcadians, who dwelt in the mountains "before there was a Moon." Plutarch also wrote of the "pre-Lunar people," and Ovid wrote that the Arcadian folk were "older than the Moon." Allusions to a time before the Moon are also found in the Scriptures (Job 25:5 and Psalm 72:5), as well as among the Cordilleras Indians of Columbia. In addi-tion, Mayan legends talk of a time when the Moon was not yet a companion of the Earth. Legends tell that Tiahuanaco, which means "city of the decaying Moon," or "city of the doomed satellite," was founded by "sky gods" from the Moon.

Velikovsky wrote that there are three theories of the origin of the Moon: (1) it originated at the same time as the Earth and is formed of substantially the same material; (2) it was formed in a different part of the solar system and was later captured by the Earth; (3) it was originally part of the Earth and was torn off, leaving the Pacific Ocean bed. Velikovsky suggested that the first and third theories are incorrect, since "in the memory of mankind, no Moon accompanied the Earth." He found it more probable that the Moon's formation occurred somewhere else, and that it was captured in Earth orbit following some type of catastrophe. This would indicate that the composition of the Moon would be quite different than that of the Earth. Whatever atmosphere it may have had ear-lier was pulled away by the Earth or dissipated in some way. Interestingly, the Finnish Runes entitled *Kalevala* recall a time when "the Moon was placed in orbit."

The journal *Science* published an article in 1970 entitled "Where Was the Moon Formed?," which concluded that the Moon was formed independently of the Earth and was later captured. The authors presumed three celestial bodies were involved, and that and these events were "followed by the dissipation of excess energy through tidal friction in a close encounter." In 1977, another arti-cle published in *The Astronomical Journal*, entitled "Stability of the Sun-Earth-Moon System," concluded that "the planetary origin and capture of the Moon by the Earth becomes a strong dynamic possibility." Studies performed in 1978, and published by Anderson (*The Moon and the Planets*) have indicated that the Moon "could not have been formed in orbit around the Earth." Furthermore, a 1978 article in *Nature* entitled "On Recent Lunar Atmosphere," indicated "strong theoretical evidence of a considerable atmosphere on the Moon during the greater part of its history." (www.velikovsky.collision.org)

Velikovsky also quoted ancient rabbinical sources which indicate that the Moon was, at one time, much brighter than it is now, and larger in appearance. It is stated that the Sun and Moon were equally bright. Both the Japanese and the American aborigines claimed the same thing. Traditions of many people the world over maintain that the Moon has lost much of its light since ancient times. Velikovsky suggested that, in order for the Sun and the Moon to be of equal bril-liance, "the Moon must have had an atmosphere with a high refracting power, or it must have been much closer to the Earth." Velikovsky proposed both assumptions. Velikovsky attributed the Moon's change in position to a cosmic catastrophe.

Because of its size and the events which accompanied the first appearance of the Moon, wrote Velikovsky, the ancients regarded the Moon "as the chief of the two luminaries." Babylonian astrologers considered the sun of smaller importance than the Moon. The Assyrians and Chaldeans referred to the time of the Moon-god as the oldest period in memory. They asserted that, "before the other planetary gods came to dominate the world ages, the Moon was a supreme deity." The Moon, since it appeared larger than the Sun, was endowed with a masculine role, and the Sun was deified as feminine. The Greeks subordinated the Sun to the Moon, as did the Vancouver Indians, the natives of Peru, and several tribes in Brazil. Velikovsky stated, "it was probably when the Moon was removed to a greater distance from the Earth and became smaller to observers on the Earth, that another name, usually feminine, came to designate the Moon in most languages."

Velikovsky also mentioned a cuneiform text describing the first appearance of the Moon, which stated: "when the gods fixed the crescent of the Moon, to cause the new Moon to shine forth, to create the month ... the new Moon, which was created in heaven with majesty, in the midst of heaven arose." (see also Rogers, *Cuneiform Parallels to the Old Testament*) Ancient traditions taught that the Adamites, or antediluvial men, established Moon worship. The Sabaeans said that Adam was born from male and female, but had "come from the Moon," and had "brought many wonders with him. (Sitchin points out that this Adam was not "The Adam" later created by the Sumerian gods.) The Adamites, Velikovsky believed, were probably *not* the first human beings to inhabit the planet Earth.

The Collision of Tiamat: Earth

Ancient Sumerian texts indicate that the Earth ("Tiamat") was struck by a large planet, which moved it into its present orbit, and created the Moon and the Asteroid Belt. In his books, *The Twelfth Planet* and *The Cosmic Code*, Zecharia Sitchin outlines this "celestial battle" as described in the Babylonian text called *Enuma elish*. The planet "Marduk" (the Sumerian "Nibiru"), as it came into the solar system on its clockwise elliptical course, struck Tiamat, which was moving in its ordained counterclockwise orbit. One of Marduk's satellites struck Tiamat first, followed by two more of Marduk's moons. Then Marduk itself, an enormous planetary body, struck Tiamat, smashing one half of the planet into pieces, which became the Earth's Moon and the "Great Band" (Asteroid Belt). The other half of the planet, which was struck by a smaller moon of Marduk, was moved into a new orbit, along with a chunk of material which became its moon. The new planet was then called "KI," meaning "cleaved one." The Earth's original moons were dispersed, many changing the direction of their orbits.

As will be discussed in more detail later, Sitchin has also written in *Divine Encounters*, the planet where God the creator, "He who is first and last," resides is referred to in the Bible as *Olamin*, the plural of *Olam*. The meaning of "olam" in the ancient world was a measure of a really long time, specifically related to the span of time between the periodic disappearance and reappearance of the planet Marduk/Nibiru on its vast 3,600 year elliptical orbit. The domain of *Olamin* was described as a kingdom which encompasses many worlds. The Dogon tribe of Mali call this planet "the egg of the world," and they say it is the origin of all things. They say of this planet that it is "the center of all things and without its movement no other star could hold its course." They say it is made of a heavy metallic compound called "sagala." The Sumerians also wrote that this planetary "god," Nibiru, "remade our solar system and remakes the Earth on its near passages."

The continent which the ancients called Pangea actually represents the pre-

The 33-mile diameter asteroid, Ida, with her one-mile wide "moon," Dactyl," recently sighted by the Hubble Space Telescope.

historic truth regarding the land formation on the Earth after it was involved in this major collision. Over time, the continents then drifted to where they are now. As Sitchin points out, Earth's largest land masses are mostly on one side and the oceans take up the rest of the planet. If you take all the water off the globe, you have a "cleaved" planet. The tale of Nibiru's Celestial Battle is actually scientifically sophisticated, and current advances in astronomy have recently corroborated certain aspects of the Sumerian cosmogony, among them the following:

- The announcement by NASA of the theory of the origin of Earth's Moon as a catastrophic collision with a "Mars-size planet." In addition, it is known that the Moon continues to move away from the Earth about two inches per year.

- Hubble's discovery of numerous planets and moons in other solar systems with highly elliptical, plunging orbits and retrograde (clockwise) orbits. The predominant motion in our solar system is prograde (counterclockwise), with the exception of Venus, Pluto, Uranus, Triton and Phobos.

- The 1994 announcement by NASA of a one-mile wide moon orbiting the 33-mile wide asteroid Ida, which follows the expectation that if a planetary body in this solar system exploded (or collided), the debris will be gravitationally bound in orbits around a primary body.

- The discovery of water, atmosphere and perhaps previous life on Mars, the Moon and Europa.

- Why there is a large gravitational "pull" on the outer planets, Uranus, Neptune and Pluto, (Uranus is tilted more than 90 degrees on its side, Pluto has a highly eccentric orbit with a large inclination), and other axial and orbital eccentricities which indicate there is another body of significant size and possibly with a heavily magnetic core beyond the outer planets.

Planet X – Myth or Fact?

In an article entitled "A History of Planet X – From the Present Day Search to the Seeding of Life on Earth" (www2.eridu.co.uk/eridu/minisites/

planetx.html), Alan Alford writes that the discovery of new planets in the last two hundred years owes more to mathematics than to bigger telescopes. Mathematical irregularities in the orbits of the outer planets, in particular, strange wobbles and gravitational anomalies noted in the orbits of Uranus, Neptune and Pluto, have prompted astronomers over the past hundred years to search for a large planetary body in the outer solar system. Based on mathematical evidence, astronomers have been so sure of the reality of this planet that they named it Planet X. The name stands for the tenth planet, as well as the mathematical symbol for an unknown quantity.

The Solar System as we know it today. We continue our search for the missing planet "X" which represents both 10th and unknown body of our solar system.

On June 17, 1982, a NASA press release from Ames Research Center officially recognized the possibility of "some kind of mystery object" beyond the outermost planets. Various press releases around this time confirmed that scientists were indeed looking for the infamous Planet X. For instance, *Astronomy* magazine published an article in December of 1981 entitled "Search for the Tenth Planet," and another article in October of 1982 entitled "Searching for a Tenth Planet."

In addition, *Newsweek* covered the story of Planet X on June 28, 1982 in an article entitled "Does the Sun Have a Dark Companion?" This article implied that the tenth planet actually orbits a two sun (binary star) system, but we cannot see the other sun because it is a "dark" star. The article stated:

A "dark companion" could produce the unseen force that seems to tug at Uranus and Neptune, speeding them up at one point in their orbits and holding them back as they pass. ... the best bet is a dark star orbiting at least 50 billion miles beyond Pluto... It is most likely either a brown dwarf, or a neutron star. Others suggest it is a tenth planet ... since a companion star would tug at the other planets, not just Uranus and Neptune.

The *Washington Post* covered the story of Planet X on the front page on December 31, 1983 called "Mystery Heavenly Body Discovered." This story reported that the Infrared Astronomical Satellite (IRAS) detected heat from an object about fifty billion miles away. A report of an interview with chief scientist Gerry Neugebauer from Jet Propulsion Laboratories appeared in the story. The article stated:

A heavenly body possibly as large as the giant planet Jupiter and possibly so close to Earth that it would be part of this solar system has been found in the direction of the constellation Orion by an orbiting telescope aboard the U.S. infrared astronomical satellite.... 'All I can tell you is that we don't know what it is,' said Gerry Neugebauer, chief IRAS scientist.

The *Post* article went on to explain that this mysterious object has never been seen by optical telescopes on Earth or in space, but was seen twice by IRAS as it scanned the northern sky between January and November of 1983. The second infrared observation of the body, which is so cold it casts no light, noted that the body appeared not to have moved in six months. This suggested that the object is not a comet, since it probably would have moved. The article also explained that the infrared telescope aboard IRAS, which is able to see very cold objects, calculated that the heavenly body was so cold that its temperature is about 459 degrees F. below zero.

Astronomers suggested it was a "giant gaseous planet, as large as Jupiter," and is so close that "it would be the nearest heavenly body to Earth beyond the outermost planet Pluto." This would make it part of our solar system. The article explained that there had been some speculation that the object "might be moving toward Earth." However, Cal Tech's Neugebauer was careful to "douse that idea with as much cold water as I can." He pronounced with certainty that this object "is not incoming mail." Neugebauer suggested it was a "dark, young galaxy that we've never been able to observe before." He explained that the 100-inch diameter telescope at Cerro del Tololo in Chile and the 200-inch telescope at Palomar Mountain in California have earmarked several nights in 1984 to search the skies for it.

The *US News World Report* on September 10, 1984 published an article called "Planet X – Is it Really Out There?" This article had the following to say about Planet X:

> Shrouded from the sun's light, mysteriously tugging at the orbits of Uranus and Neptune, is an unseen force that astronomers suspect may be Planet X – a 10th resident of the Earth's celestial neighborhood. Last year, the infrared astronomical satellite (IRAS), circling in a polar orbit 560 miles from the Earth, detected heat from an object about 50 billion miles away that is now the subject of intense speculation.

The article went on to say that scientists are *hopeful* that the Pioneer 10 and 11 space probes will locate the object, which they noted was possibly a "brown dwarf," a protostar that never got hot enough to become a star. Others astronomers, however, argue that the object is a "dark, gaseous mass that is slowly evolving into a planet." The article quoted Neugebauer as stating: "If we can show that our solar system is still creating planets, we'll know that it's happening around other stars, too."

Contrary to this information, according to the ancient Sumerian texts, our solar system is not still creating planets, and this planet has been with us all along. It would appear that the media spin being put on Planet X is a clear attempt to call it something else besides "incoming mail."

The media was quiet on the subject of Planet X for the next four years. Finally, an article by R. Harrington in *The Astronomical Journal* dated October 1988 supplied the details of continuing mathematical modeling of this planet. The article suggested the mysterious planet was three to four times the size of Earth, and its position was three times further from the Sun than Pluto. Mathematical modeling also suggested that Planet X had an extreme elliptical orbit of 30 degrees.

A NASA/ARC press release published in *Newsweek* on July 13, 1987 disclosed that "an eccentric 10th planet may – or may not – be orbiting the Sun." The article stated that NASA research scientist John Anderson "has a hunch Planet X is out there, though nowhere near the other nine." The article concluded, "if he is right, two of the most intriguing puzzles of space science might be solved: what caused mysterious irregularities in the orbits of Uranus and Neptune during the 19th Century? And what killed off the dinosaurs 26 million years ago."

The reference to killing off the dinosaurs seems to indicate that something more is known about this planet than NASA is letting on. Is this in fact a reference to the planet Nibiru of ancient infamy, the planet which, according to ancient sources, did in fact strike the Earth (Tiamat) in ancient times, and gave us a moon for a thank you? Regardless of what NASA really knows, silence is golden, and, for the most part, mum was the word on the issue of Planet X during the 1990s. Instead, scientific journals began to debunk the issue of a tenth solar system planet.

The issue of Planet X has now become entirely befuddled with recent discoveries of 18 nearby stars with Jupiter-class planets orbiting them (last count as of October, 1999). (see astron.berkeley.edu) Locating any new developments regarding a tenth solar system planet is akin to finding a needle in a haystack, since it's difficult to know which "new planet" they are talking about. In the 1990s, news stories began to dwell on "runaway planets" and "rebel planets" discovered in other solar systems.

For instance, an October 23, 1996 Associated Press article entitled "New rebel planet found outside the solar system," disclosed the following:

> A new planet that breaks all the rules about how and where planets form has been identified in orbit of a twin star about 70 light years from Earth in a constellation commonly known as the Northern Cross. The new planet has a roller coaster like orbit that swoops down close to its central star and then swings far out into frigid fringes, following a strange egg-shaped orbit that is unlike that of any other known planet.

Thus, the issue of Planet X became lost in the information shuffle. Due to the Hubble space probe, many distant galaxies, stars, planets and brown dwarfs are suddenly being discovered all at once. This situation affords NASA the opportunity to attempt to avoid the societal chaos that will surely ensue once everybody realizes that a sizable planet somewhere out there is actually destined to circle our own sun in the next couple of years. After all, who would bother to go to work, go to school, pay their bills, play the stock market, be a loyal, tax-paying John and Jane "Doe," if they thought the end of civilization was nigh? There would be some who would just "charge it" to the hilt and party 'til the cows come home (that would be me), and others would go berserk in other ways.

Contrary to what happens in Hollywood movies, where the government comes clean about comets (actually, in *Deep Impact* they waited almost a year after official knowledge to announce it to the public), wouldn't it rather make sense that such an announcement would not be made public, at least too far in advance of D-day? Tom van Flandern of the U.S. Naval Observatory, who was quoted in the 1982 *Newsweek* article, admits that a tenth planet is possible, but argues that it would have to be so huge that it *should have been observed by now*. We must admit that there is a serious possibility that more details regarding Planet X are in fact known, but we are not being informed. Is there a major cover-up of Planet X?

This possibility was actually discussed in an article in the *CCNet Digest* on May 18, 1998 entitled "The Secrets of Asteroid Peril, British Media Smells a Rat." This article claimed that a report in the London *Daily Mail* accused US astronomers of trying to cover up scientific data until NASA has a chance to look at the information. The report indicated that "earth-shattering" information, such as the discovery of incoming comets or asteroids, has been ordered to go through NASA first. The *Daily Mail* report of May 15, 1998 entitled "Delayed Impact, or the Secrets of Asteroid Peril," specifically stated the following:

> If a giant asteroid is hurtling in the general direction of our planet, we will be the last to know about it. For astronomers have decided that the news would be too earth shattering for ordinary mortals to handle, and would likely cause widespread panic. In a week that sees the release of the film *Deep Impact*, a fictional account of just such a catastrophe, astronomers funded by the American space agency NASA have now agreed to keep asteroid and comet

discoveries to themselves for 48 hours while more detailed calculations are made. The findings would then go to NASA, which would wait another 24 hours before going public.

A Los Angeles Associated Press article on May 19, 1998 also reported on this 72 hour delay rule imposed by NASA on announcements to the public of asteroids or comets by astronomers. The report indicated that these measures were undertaken to avoid the "doomsday alert" which occurred in March of 1998 with regard to an asteroid which was due to collide with Earth in 2028, and which was "soon found to be a mistake." These new procedures, the article stated, are "not an attempt to hide anything but to make sure the information is accurate." How can we be sure that if the information is found to be accurate, NASA will not (a) find it to be a "mistake," or (b) withhold the message for a longer period of time?

In another interesting press release dated June 5, 1997, the Associated Press discussed a "Tiny Planet Discovered Beyond Pluto." The article suggested that this object was in fact part of our solar system. This 1997 press release stated:

> Astronomers have found an icy miniplanet that orbits the sun well beyond Pluto, providing evidence that the solar system extends much farther than was once thought. At its most distant, it wanders three times farther from the sun than Pluto, tracing a looping, oblong path into an astronomical terra incognito. The discovery of the 300-mile-across object has extended the known edge of the solar system's Kuiper Belt by at least 9.35 billion miles.

This news report is interesting in several ways. It indicates an extension to the solar system, it announces that planets can travel in highly elliptical orbits, and, I am venturing a guess here, it is the first announcement of the sighting of one of the small moons of Planet X.

In February, 1999, J.B. Murray presented a paper to the Royal Astronomical Society entitled "Arguments for the Presence of a Distant Large Undiscovered Solar System Planet." (see www.blackwell-synergy.com) Murray's paper explored various explanations for what he called a "non-random clustering" of "long-period comets," which his research concluded are "aligned along a great circle inclined to both the elliptic and the Galactic plane." His paper examined the possibility that this non-random clustering was due to "orbital perturbations by an undiscovered object orbiting within the … distances of 30,000 to 50,000 au from the Sun." Murray's mathematical modeling predicted that the object would have a retrograde orbit inclined at 120 degrees. He noted that:

> Such a distant object would almost certainly not remain bound for the age of the Solar system, and recent capture into the present orbit, although also of low probability, remains the least unlikely origin for this hypothetical planet.

An article published in *The Economist* on October 16, 1999 entitled "X Marks the Spot" made the following dubious remark, which pretty much sums up the status of Planet X:

> Two astronomers claim to have found evidence of a tenth planet orbiting the sun. Or another star that didn't quite make it. Or at least something.

The article went on to explain that scientists have actually been looking for this mysterious and gigantic solar system planet since 1846. The reason is that anomalies in the orbits of Uranus and Neptune cannot be caused by the gravitational influence of Pluto, since it is too small. There has to be something much larger out there, astronomers have presumed for over a hundred years. However, as disclosed in "X Marks the Spot," Drs. Murray and Matese, after looking at the orbits of approximately 300 long-period comets, have separately concluded that too many of them are coming from the same regions of space. They suggest that the galaxy's "tidal wobble" is "being modulated by the gravity of something big within the Oort Cloud itself."

The new object must be very faint, these astronomers suggest, or it would have been spotted. This means it's not a star. They predict that the object is three times the size of Jupiter. They also suggest that the object is not a "proper planet," because, take note: "its orbit appears to run *in the opposite direction from those of the nine known planets*." This is another factoid we can add to the above "anomalous" findings regarding the existence of Marduk/Nibiru. As the *Enuma elish* tells us, the planet Marduk entered the solar system on its "clockwise, elliptical orbit," and struck the Earth, which was moving in its "ordained counterclockwise orbit." So the planet we are looking for will have an orbit which runs in the opposite direction from those of the nine known planets in our solar system. (Chalk another one up for Sitchin!)

According to the *Economist* article, the orbit of this object also appears to be so unstable that it is unlikely that it could have maintained this orbit for the 4.5 billion year history of the solar system. Therefore, these astronomers suggest, it is more likely that the object is an extra-solar body that was captured by our sun's gravity only "recently," in astronomical terms. (Otherwise, it would have wreaked virtual havoc throughout the history of the solar system. Of course, we are presuming that our neck of the galactic woods is serene because we've been left here in peace to "evolve" for so long that Planet X *would* have to be a relative newcomer to fit into this preconceived scenario.) In any case, these astronomers predict we should be detecting Planet X soon, even if we have to wait for the next generation of infra-red space telescopes.

It's interesting to note that many of the newly discovered extra-solar planets recently discovered have extremely "eccentric" orbits. It is suggested in "New Discoveries, A Planetary Mystery" (see astron.berkeley.edu/) that circular orbits may actually not be the rule in nearby solar systems. Nine of the extra-solar objects recently discovered travel in unusually elliptical orbits, "several of them plunging in relatively close to their stars and then swinging far out again." Several of these planets may be three times as massive as Jupiter, and one is estimated at eleven times the mass of Jupiter. If they weren't so big, we wouldn't be able to detect them, since only massive objects cause mathematically detectable gravitational anomalies.

What these new discoveries are teaching us is that our own solar system may be very unlike others. Even considering our catastrophic past, we may inhabit a relatively peaceful end of the cosmos, as far as galaxies go. As Dr. Geoffrey Marcy explains, powerful gravitational forces exerted over smaller planets by huge planets or passing stars are capable of upsetting their orbits. Likewise, two or more huge planets orbiting in close proximity can generate a "gravitational slingshot" effect, which could fling one of the objects off on an elongated orbit into the inner planetary system, while the other could at the same time fly off in the other direction, perhaps escaping into interstellar space. Or, in another possible scenario, the planet's own star (sun) could be part of a binary star system, locked in a gravitational embrace. Such a companion star would be very capable of exerting a catastrophic influence on the orbits of nearby planets.

As Dr. Marcy writes: "We are realizing that most of the Jupiter-like planets far from their stars tool around in elliptical orbits, not circular orbits, which are the rule in our solar system." Why our largest planets remain in circular orbits is, therefore, a mystery. As Dr. Marcy notes:

> Jupiter sized bodies plunging toward and away from their stars are likely to sweep aside smaller worlds, sending them crashing into their star or flying out of orbit into interstellar space. Current technology is incapable of detecting Earth-size planets around other stars, but they almost certainly could not exist near their star's warmth in a system so unsettled by large planets in wrecking orbits.

A huge planet on an elliptical orbit would probably scatter or destroy smaller planets as it crosses their paths time and again. As Dr. Marcy writes, "if our Jupiter were in an eccentric orbit, the Earth and Mars would likely be gravitationally scattered out of the solar system." Therefore, our very existence, he writes, is dependent upon both Jupiter and Earth being in mutually stable and circular orbits. Lucky for us, we have no big bullies, no freewheeling wrecking balls, in our solar system. Or do we? Enter Planet X.

Interestingly, astronomer Tom Van Flandern wonders if Planet X has actually exploded. In his book, *Dark Matter, Missing Planets and New Comets*, Van Flandern wonders if this planet was responsible for disruption to the moons of Neptune. In his 1995 paper entitled "Origins of Trans-Neptunian Asteroids," Van Flandern notes that a new asteroid belt has been discovered beyond Neptune, suggesting that Planet X may have exploded. (On the other hand, why wouldn't this be the result of one of Planet X's fly-bys?)

In his book *The Phoenix Solution*, Alan Alford asks, if Planet X has exploded, why do the Mesopotamian texts imply that it was seen by the Sumerians during the 4th-3rd millennia BC? Alford suggests that the planet Nibiru was not actually seen, and is a "metaphysical" planet. He believes that the descent and ascent of the Egyptian 'gods' was metaphysical, and that the Egyptian religion was "an exploded planet cult." The story of Osiris was based on this exploded planet; the deity Osiris had been mythologically "dismembered" and his body distributed among the other gods in the solar system. The resurrection of Osiris [continued in the Jesus myth] was based on his metaphysical journey to the stars, which restored the exploded planets to their primeval and pristine form; essentially a metaphysical trip back to "The First Time."

Alford charges that Sitchin has misinterpreted the gods as interplanetary astronauts, and that Planet X actually did not come into the Earth vicinity, but was essentially a "play" or metaphysical replay of an even more ancient event. He also tries to incorrectly establish a Darwinian evolutionary trajectory that implies human evolution was "quickened" by this explosion of planets. This is in fact an anti-Darwinian thesis, but we will get to that later.

More importantly at this juncture, do I understand correctly that what IRAS spotted in 1983 was a metaphysical planet? So that means that what NASA officials mistook as a heavenly body in the direction of the constellation Orion, which they claimed was possibly as large as Jupiter, and close enough to Earth to be part of this solar system, was a mythological planet? And so the mysterious heavenly orb which mathematicians now predict may be three times the size of Jupiter, with a highly elliptical orbit that runs opposite the other nine planets in the solar system, and a retrograde orbit inclined at 120 degrees, is just the replay of an ancient event in the solar system? As the famous line in Oliver Stone's *JFK* goes, "people, we're through the looking glass here."

The Hyper-D Physics Connection

In his paper entitled "Hubble's New Runaway Planet," found at www.enterprisemission.com, Richard Hoagland explains the forgotten 19th Century scientific theories of Hyperdimensional Physics. These theories were initially proposed over a hundred years ago by various scientists, and were updated in this Century by Faraday and Maxwell. In a nutshell, what hyperdimensional physics tells us is that three dimensional physical phenomena, perhaps including human consciousness itself, are dependent upon higher dimensional realities for their existence.

Hyper-D physics hypothesizes that the basic laws of the three-dimensional world are united in four dimensional space, but merely "look different because of the resulting crumpled geometry of 3-D reality." This is a major break with

Newtonian laws of "action at a distance." Instead it proposes that such "forces" are a "direct result of objects moving through 3-D space geometry … distorted by the intruding geometry of 4-D space." In essence, the laws of nature become simpler in higher dimensions because of the effect of geometric spatial distortions in the 3-D world.

The implications stemming from this model—the dynamic hyperspace foundations of reality—completely demolish our current view of reality. For, according to Hoagland's research, the "tail wags the dog." Hyper-D physics is the only model that can explain how the planets are completely capable of exerting a determinant influence on the Sun and on each other through their disproportionate ratio of total solar system angular momentum. Suffice to say, according to Hoagland, we are simply entering once again, after 13,000 years (which incidentally is equal to the Mayan "Baktun" cycle) "a phase of this recurring, grand solar system cycle of renewed hyperdimensional restructuring of reality." And, as Hoagland notes, this may even be due to the presence of a particularly enormous solar system body which has just recently come into view, thanks to the Hubble telescope. As Hoagland notes:

> Described first by Kepler as his "Third Law," the farther out a planet orbits from its star the longer is its period of revolution. Since we're talking about possible additional planets driving this Hyperdimensional Solar Process, and planets that must be hundreds of times farther from the Sun than Earth (Pluto is "only" 40 times its distance, and orbits in "only" about 250 years), the "years" of these extremely distant worlds could equal thousands if not tens of thousands of Earth years, depending on their orbits. Because of these immense orbital periods, the cycles of solar energy production driven by their combined angular momentum will be v-e-r-y long indeed.

As a matter of fact, long enough to throw off Earth time. As Hoagland notes, something strange is indeed happening in the solar system. One of the effects is that the official Cesium Atomic Clock has had to be adjusted by over 20 seconds in the last two decades; approximately a full second every six months. The official atomic clock is supposedly theoretically stable to plus or minus one second in a million years. To Hoagland, this suggests that a progressive phase-shift is occurring between the Earth's rotation and the atomic-level constants that govern the quantum standards of the atomic clock.

The growing difference between dynamic time and atomic time is compelling evidence of large scale changes in the hyperspatial physics of the solar system. It is possible, Hoagland explains, that hyperspatial stresses are due to the progressive orbital movement of as-yet undiscovered outer planets, which are causing increasing scalar potential changes across the entire solar system at the atomic level. The reason the clocks are changing, however, is because the Earth's rotation is changing. It is essentially slowing down due to changing phase relationships between undiscovered outer planets in the outer solar system! There can be no doubt that something is out there! And this something is quite capable of "making and remaking" the Earth, including seriously affecting the orbital rotation time of the planet.

It's also interesting to note in this regard that if orbital revolution is directly related to the passage of "time," does this affect the aging process of the life forms on the planet? Could the presence of Marduk/Planet X be a possible explanation for the longer ages of the Biblical patriarchs? Also, does "making and remaking the Earth" refer to a higher vibrational consciousness referred to as 4th dimensional consciousness? Is there then a relationship between the end of the Mayan Baktun cycle and the coming of Planet X? In any case, these bizarre "new age" concepts seem to be making more and more scientific sense.

Such a distant and large solar system body will also have a huge effect on total solar hyperdimensional energy generation, Hoagland notes. He explains, unknown "worlds," or planets, can produce long-term cyclical changes in solar output lasting thousands of years [i.e. a "World Age"?]. In fact, such a geometrical relationship has already been established between the changing positions of Jupiter and Saturn relative to each other. Hoagland wonders:

> If the known changes in solar output are due to hyperdimensional effects of the largest known planets, what of the magnitude of "aether stress" produced by our proposed "new planets" with angular momentum contributions hundreds of times greater? The long-term cyclic increase in solar energy created by those cyclic phasings (yet to be experienced in recorded history) could measure as much as several percent above current solar output. This is more than enough additional energy, even without Mankind's current addition of significant greenhouse gases to the atmosphere, to trigger profound, millennia-long climatic changes here on Earth.

Can human events be "influenced" by the planets? If so, to what degree? Can this influence be considered "intelligence"? As Joseph Mason has noted with regard to the Chakra System and Menorrah-shaped crop circles that are now appearing in the U.K., these magnificent forms appear to be the intelligent communiques of a hyperdimensional or 4th dimensional conduit. Is this what could be meant by "makes and remakes the Earth on its near passage?"

If the hyperdimensional physics model is correct, Hoagland writes, we are now well into a new long-term cyclic solar period. Yet, can we apply this distant effect to particular human behaviors? We are certainly a long way from understanding these concepts, if only because the science behind them is so profoundly different from our accepted view of causality. The intelligent and purposeful cyclical planetary "fates" have never been so far from human consciousness as they are now. As we shall see in the upcoming chapters, the ancients believed that the planets had great influence on the Earth and upon human morality and endeavors.

In light of information regarding hyperspace reality, it appears we have come full circle. Human knowledge has a way of biting itself on the ass. For, as the pioneering work of 19th Century physicists tells us: "The existence of unseen hyperspatial realities, through information transfer between dimensions, are the literal 'foundation substrate' maintaining the reality of everything in this dimension." People, we're through the looking glass here.

How are we to assimilate this information? Is it insanity to believe that our blindness to human fate has been orchestrated by geo-magnetic planetary forces? As Velikovsky has noted, could our forgetfulness of our origins be attributable to ancient planetary magnetospheric anomalies? Are we destined to have to work so hard to understand our human situation, or is it simply part of the memory lapse descriptive of the World Age that we are now moving out of?

If I were the only one asking these questions, I'd think about checking myself in. But if the exceptional synchronicitous knowledge coming at us at this time constitutes insanity—as Hoagland notes, the message in this "signal" is increasing in strength and frequency—then there are scores of people who ought to be medicated and wrapped in white jackets. On the contrary, insanity in such large numbers can only mean that the population is waking up! Good morning Earth People! Gather round and listen to what's been going on while you were asleep!

In assimilating this information into our world view, we have to realize that our science is missing large areas of information with regard to geometric interplanetary relationships in our solar system's cosmic dance, and the peculiar physics of a shared outer solar system planet or brown dwarf scenario, and the connection this might have on both Earth's climate and rotation, and also, as strange as it may sound, on space-time reality *and* human consciousness.

How much meaning can we, in our modern secular world view, place on the distant effects of interplanetary hyperspace gematria? In our blindness to important scientific discoveries of only a hundred years ago, there is no context to explain these compelling synchronicities. Conceptually speaking, we have nowhere to put this bizarre information. But we need to become aware that a Glass Bead Game of enormous importance is beginning to take shape, and all hands are needed on deck to decipher the mathematical codes of this information. We are being spoken to in a forgotten ancient language, and we've got to wake up fast to the meaning of the message.

If the "supernatural" underpinnings of most of this information is too far-fetched for some readers, we must realize that the supernatural becomes a natural matter when we assume both quantum and hyperdimensional physics to be descriptive of reality. These ideas will be covered in more detail in my next book, *Phenomenal World*. Nonetheless, if this is still too far to go, it is certainly within the realm of possibility, in a more mundane sense, to apply our current understanding of cause and effect to a particular human behavior of obvious urgency: that is, the building of the International Space Station. What is the urgency behind this "Alternative 3" maneuver? Let's continue with the exploration of this matter.

A Moon For a Thank You

It is now almost certain that the Moon was once part of the Earth and, according to ancient legends, there *was* a time when the Earth was without its present companion, or, perhaps, had a different, larger companion. But, we have to ask ourselves, how are "pre-scientific" people able to recall a time without the Moon, or a time when the Moon was "placed in orbit"?

The reason Velikovsky did not concur with the theory that the Moon had once been part of the Earth was due to the fact that Earth people were aware of a time when the Earth was without the Moon. Placing this event within humankind's memory does not make sense in the context of this very ancient event. The only way we can make sense of it is to allow a very ancient origin of humankind on this planet prior to the collision which split off a huge section of the planet. But the human evolutionary paradigm will not allow this. (We will explore this matter in a later chapter.)

In addition, if people *were* here, how could they have survived such a calamity? Could it be that those who told the stories simply heard them from extraterrestrial travelers who "witnessed" the event from a distance? Are we essentially the progeny of these space travelers? Was the genetic stock of the human race saved by space travelers in order to avert total annihilation of the species? If we take into account how many potential planetary catastrophes have occurred, could this be an ongoing "assignment" of certain groups of space traveling genetic scientists, or their self-replicating programmed probes?

It is also interesting that many ancient stories report that the Moon was once larger (or closer) than it is presently. How can anyone have known this if there was nobody here to witness it? Could it be that after the Earth was struck, and when the Earth first captured this chunk of material as a satellite, the Moon was much closer to the Earth and it was slowly moved away by the Sun's gravitational pull? This could obviously have occurred. Another possible explanation could be that these ancient reports are references to another moon entirely. As Sitchin explains, Tiamat lost its original moon when it was struck by Marduk. Also, might the large planetary sphere, Marduk itself, have been mistaken for a moon when it was in the Earth vicinity?

There is also a possibility that the Moon was "deliberately" moved. As Sitchin reports, the Sumerians believe that the planet Nibiru "makes and

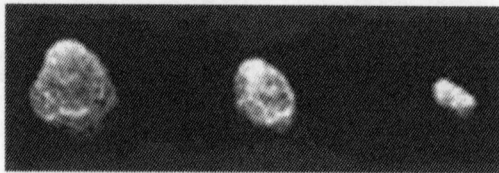

Hyperion, a moon of Saturn (220 mi by 130 mi and shaped like "a hamburger") is probably not in a gravitationally stable position. According to NASA, it is possible that a meteorite impact moved it out of position and that it will swing back gradually. Taken by Voyager 2, Aug 25, 1981.

remakes the Earth on its near passage." This could simply mean that Nibiru demolishes Tiamat in its catastrophic Shakespearean re-play, and leaves the results to cosmic fate. Or it could mean that something is done to alter fate, or perhaps restore the rightful positions of the play's "stars" afterward in an attempt to minimize the catastrophic results and, under the circumstances, to create "the best of all possible worlds."

Interestingly, an article on the Enterprise Mission web site entitled "Oh My God, They've Killed Soho" suggests that a "hyperdimensional" vehicle would be capable of "nudging" and gently pushing an asteroid into "a precise orbit of virtually any configuration." This paper suggests such technology is implicit in the "HyperD" model, and that if we had such technology we could use it to avert comets and asteroids coming our way. This article asserts that our secret space program has long used suppressed physics models to produce "scalar" gravity control technologies, and that such vehicles would be able to bump a comet or asteroid into a specific orbit with no effort whatsoever. As a matter of fact, enormous hyperdimensional vehicles have been spotted on two videos taken by NASA on shuttle missions. In both cases, the cameras record powered vehicles flitting about in low-Earth orbit in full sight of Houston's cameras.

The idea of a small moon being moved into another orbit using hyperdimensional physics is not as ludicrous as it sounds, given the large dimensions of some of the flying disks reported and the small size of some of the small moons of Saturn, for instance. Small moons are usually called "icy moons" because they are so lightweight that it is *surmised* they have any icy core rather than a metal core. For instance, several of Saturn's moons are only about the size of a large city or small county. The Moon Helene is only ten miles across. The Moon Hyperion is about 117 miles across and is shaped like a "soda pop can." The "twin" moons Epimetheus and Janus orbit Saturn together and are only 33 miles apart. Epimetheus is about 45 miles across and Janus is about 75 miles across. Ganymede and Enceladus have "grooves" which extend many kilometers on the surface, and do not have craters. Enceladus is about 330 miles across, and is "highly-reflective." Uranus's Moon Miranda is only 200 miles across and has low density. Could some of these strange little moons be "terra-formed" bio-spheres? Or, could civilizations be living *inside* them, a la *The First Men In the Moon*?

According to an article I recently discovered by Chris Boyce entitled *The Logical Contact*, (www.et-presence.ndirect.co.uk/Articles), intelligent interstellar communications are likely to be contained within a von Nuemann (vN) probe-like network, "rather than blasted out to the whole universe." This would mean that, contrary to SETI's belief, we should expect communiques to be coming in locally rather than from outside of the solar system. Boyce writes: "any base they have for studying our planet is unlikely to be sited further away than the Moon and more likely to have been placed somewhere right here on Earth many millions of years ago." Boyce also suggests that it is unlikely that such an intelligence network would be awaiting our call, since "the ubiquitous vN" will already know all about us. As a matter of fact, he writes, if they really wanted us to "sit up and take notice" they would be more likely to leave "a local marker somewhere in the Solar System to prompt our curiosity and nudge our thinking."

This suggestion would seem to apply to ancient artifacts noted on the Moon and Mars, as well as to the incredibly intricate and apparently remotely-microwaved crop circle designs that are appearing non-stop in the U.K. With regard to radio communiques, Boyce writes that Long Delayed Echoes (LDEs) have been observed in the solar system for over 70 years. One hypothesis is that LDEs are evidence of interstellar probes in our solar system.

In short, the idea that advanced extraterrestrial space travelers could apply hyperdimensional technology to engineer orbital catastrophic management projects is actually within the realm of possibility. After all, we must consider the type of technology that would be important to an intelligent civilization living on what constitutes a runaway wrecking ball. Might the technology of such a society be based on the altruistic behavior of running ahead and moving objects out of its path, and running back to tidy up the bowling alley afterward? Might the technology also be based on placing and replacing the genetic stock of life forms on planets on its course? Even more far-fetched, could hyperdimensional technology use light and sound vibrational levers to lift the Earth globe right out of the 3rd dimension in order to avoid a head-on collision with the oncoming planet?

If the mysterious Planet X exists, it makes sense that the society calling it home would post intelligent self-replicating probes throughout solar system neighborhoods bordering their own as scientific outposts. As Boyce suggests, if ETI exists, their probes are certainly already present in our own solar system, "perhaps with a presence here on Earth or close by." As we will get to in great detail later in this book, Boyce also suggests, "elements within the vN presence would probably be capable of fabricating somas, near perfect artificial likenesses of biological forms." Let's continue our investigation of the proof of space travelers in our solar system.

The Ballrooms of Mars

The present desolation of the Moon and Mars and other celestial bodies, wrote Velikovsky, does not imply that in the past they were desolate. Velikovsky believed we had the testimony of our ancestors, supported by modern observations, that Mars and the Moon were engaged in near-collisions only a few thousand years ago. He wrote: "It is not excluded that under conditions prevailing on their surfaces prior to these events, life could have developed there or elsewhere in the solar system to an advanced stage."

In 1965, the unmanned robot probe Mariner 4, followed by Mariner 6 and 7 in 1969, beamed back to Earth pictures of what appeared to be a dead Martian landscape. However, in 1971, Mariner 9 returned cleaner, more detailed, and much more interesting frames. In his book, *Architects of the Underworld*, Bruce Rux describes these pictures. In addition to pictures of pyramidal objects in "rhombus formation," another picture, frame 4212-15 was dubbed "Inca City" for its resemblance to ancient ruins. Frame 4209-75 showed a formation which NASA described as "unusual indentations with radial arms protruding from a central hub." To some, this anomaly looked an awful lot like an airport terminal, but to NASA it was just "an unusual formation."

In 1976, the Viking 1 orbiter took a picture of the Cydonia region from 1,000 feet above the Martian surface. Picture frame 35A72 was a picture of a formation resembling a human face of extraordinary proportions: one mile across, two miles long, and a half-mile high. NASA assumed it was a trick of lighting, and fully expected the face to go away the next time around. Thirty five days later, Viking 1 relayed a second picture of the face. This time, it was even clearer. The now infamous "Face on Mars" is geometrically aligned with several pyramids in the locale via a geometrical constant Hoagland calls "the tetrahedral message of Cydonia."

It turns out that the pyramids and Sphinx in Egypt (the Giza complex) are aligned on exactly the same principle of tetrahedral geometry. So are the pyramids in South America. The astrological alignment of certain features of the Giza plateau indicate that Sitchin may be correct in asserting this was once Mission Control Center. Ironically, Barrow & Tipler explain how a messenger probe could construct an artifact in the solar system of the species to be contacted; an artifact "so noticeable that it could not possibly be overlooked." They muse that "once the existence of the probe has been noted by the species to be contacted, information exchange can begin in a huge variety of ways."

Yet, it is obvious that these musings are purely theoretical, since the potential extraterrestrial artifacts just discussed, along with other anomalous writings on the wall, have accomplished nothing but information coagulation between Earth powers and Earth peoples. Could it be that such information exchange has gone on privately between Earth powers and extraterrestrials?

Indeed, there is profound mathematical evidence of a past ETI civilization on Mars, photographic and telescopic evidence of colonizing activities on the Moon, and perhaps evidence that the moons of Mars are huge artificial satellites; all of which evidence NASA has decided does *not warrant* closer inspection. At least *our* inspection. Behind closed doors, however, you can be sure the powers-that-be are indeed very interested in these anthropomorphic anomalies.

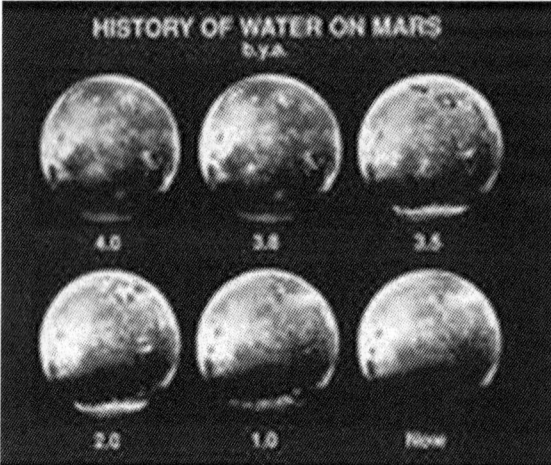

History of Water on Mars. The latest solar system discoveries are: the discovery of water on Mars, the Moon and Europa; as well as possibly beneath the surface of Jupiter's moons, Callisto and Ganymede, and Saturn's moon, Titan. In addition, a meteorite which fell in Texas in 1998, contained water. Hubble has detected traces of molecular oxygen in the atmospheres of Mars, Venus, Ganymede, and Europa. Nitrogen/methane was detected on Titan and Triton (Neptune's largest moon), ozone was detected on Ganymede, sulfuric acid was detected on Europa, and sulfur dioxide on Jupiter's moon, Io.

The scenario offered in apparent jest by the Space Travel Argument's proponents might very well depict reality once we remove the blinders of the reality engineers. There is simply no basis for the Space Travel Argument's blatant assertion that ETI have never visited Earth or any planet within our solar system. Scientists could not possibly be privy to the truth of this matter; therefore, the Space Travel Argument is not based on any known truth. Instead, since it defines what it considers proper evidence, it seems as though it has been contrived in order to confine the parameters of human thought, and to hold in place an entirely engineered reality system which tells us that 2+2=5. Let's continue our exploration of the facts of this matter.

Chapter Three

Who Constructs Reality?

Orwellian Paradigm Control: "2+2=5"

Americans are peculiarly vulnerable to stagecrafted entertainments. Americans are the most manipulated people in the world.

Wilson Bryan Key
The Age of Manipulation

Officially, our government *has* to know much more about the unusual features of the Moon than it has openly admitted: the fact that large objects move around on it, that lights have been seen moving on it, and vehicle tracks as well as huge artifacts have appeared in satellite pictures. Isn't it also curious that the moons of Mars weren't seen before 1877 even though there were people studying Mars all along? Can these moons actually be terra-formed bio-dome satellites pushed into their orbits by hyperdimensional vehicles only a little over a hundred years ago as the space stations of advanced space travelers?

In his book, *Penetration*, Ingo Swann offers evidence that our Moon is a man-made satellite and is quite possibly occupied by extraterrestrials. In light of evidence that it is actually a piece of the Earth, I doubt the assertion that the Earth's moon is man-made, but there's reason to believe it may be occupied. He suggests that the American government and higher echelons at NASA have long had evidence of the Moon's occupation, but we are "entrained" not to think about it. Let's explore the ramifications of this shocking assertion.

Swann's interest also involves the subject of telepathy. Swann is a talented psychic and was involved for many years in the government remote viewing program at Stanford Research Institute. What is the deep black government's obsession with human telepathy? Does it really pertain to a Soviet psychic Cold War, or should we be thinking in terms of Ron Reagan's peculiar statement (and perhaps veiled allusion) to the effect that a visit by an ET race would cause us to think of the Russians as our allies? Are deep black agencies attempting to find out if humans are up to the task of a telepathic War of the Worlds? Are ETs here on Earth among us? Is our full conscious awareness shrouded by some kind of remotely-caused "stupor" when it comes to realizing that certain bodies in our solar system may be occupied, or may even be man-made constructs? How is it that we are unable to make this obvious connection?

Penetration, is one of those stories you have to put down in your lap every so often, close your eyes and *hope* it's 90% fictional. Even the well-stacked female squeezing the artichokes in the produce section ain't no lady. She's an extraterrestrial. It's not that these ideas haven't waltzed the fringes before. It's just that nobody has been allowed to discuss Black-Ops programs dealing with earthbound ETs without putting it in fictional format. It's curious that Swann is being allowed to do this now. Is it part of an ETI "Acclimation Program" believed to be in effect?

Penetration reveals the story about a "deep black" agency which utilized Swann's psychic talents in order to "remote view" man-made artifacts and humanoid occupants on the Moon. Two very strange "twins" would meet him and take him to secret meetings with a "Mr. Axelrod." The "Axelrod Affair," which took place in 1975, was so undercover that there was to be only a verbal secrecy agreement lasting ten years. There would be no paper trail. Ingo was taken to an underground facility outside of the D.C. Beltway where his remote viewing sessions of the Moon took place. He was also taken to Alaska where he viewed an enormous "growing" diamond-shaped UFO rising out of a lake, which he describes as an "appearance," rather than an "object." He speculated that this thing was some kind of remote-controlled water vessel performing collection duties, since it seemed to expand as though it were taking in water.

The focus of the first part of *Penetration* is on these bizarre "twins," who seemed to be telepathic. Somehow they even knew the combination to Swann's locked office at SRI. As Swann tells it, he would see these twins everywhere, even at the grocery store following the well-stacked extraterrestrial. This makes him wonder who Axelrod is really working for: the CIA, KGB, Mossad, M-5, military intelligence, or ... "worst of all was the speculation that they, themselves, might be extraterrestrial." He wonders, could ET troops be fighting a war on our very own turf? We will come back to this discussion shortly.

A science fiction film which touches on these subjects is *Dark City*. In this film, ETs shut down the city every night at midnight to implant memories into humans. Each of the humans has only just become that bundle of memories the night before, or perhaps up to a few weeks before. No events are real, they are only transplanted memories. Someone else may have had those memories before they became yours. The trains travel only to the limits of the city. It is all a construct sitting out on a manmade platform. But one human discovers he has the same powers of telepathy that the ET overlords have. This is his ticket to ride: the ticket to change reality. This bizarre mind control scenario is not what I am suggesting as the present human condition, but I think it symbolizes a potential future human condition if we don't wake up to these subtle manipulations of our minds. But how would the initial phase of such manipulations begin? It may begin with a "potion," a psychic recipe aimed at the severing of the "quantum self."

In his book, *The Self-Aware Universe,* Goswami explains that the true nature of human consciousness, or the "quantum self," lies at the transcendent level. As a general rule, the ego takes over and responds in pre-ordained ways. But in novel situations, the human mind is capable of kicking up and out of the system to access a level of consciousness that is collective and universal. As a truly non-local consciousness, he explains, we operate from outside the system. This realm of consciousness has also been called the Universal Akashic Records, and has been described as an energy field or vibratory record of thoughts and events stored in the collective well of Mind-at-Large. Goswami asks those who do not believe in a "quantum self" to imagine what life would be like if there were a potion that could sever the quantum self. This is a scary thought indeed. Yet, perhaps there is a "potion" at work which is capable of severing the quantum self. A potion usually implies something "magical" or "ritualistic." In everyday terms, this might translate as the daily insidious ritual of media information management, and the potion itself might be the ritual act of planetary tricksters who delude our senses from the minute we wake up until the minute we go to sleep.

If we approach it as an enormous Glass Bead Game, we might see connections between the parallel information being offered by various sources. At this time, every hand is needed on deck.

In his book *The Age of Manipulation*, Wilson Bryan Key reminds us that a perceptual construction is a tentative view of a changing reality. Fixed opinions tend to congeal "reality" into a choice-related phenomenon, concealing the variety of options that are actually available. The same process of perceptual priming can describe both advertising hype and media propaganda. News media is just as "merchandised" as are the products we are urged to buy day in and day out. The daily news is just as "peddled" as product information is "propagandized." In the end, we are primed to accept a symbolic reality which tells us the Big Lie: that we think for ourselves. However, in the Western world, the military-industrial-capitalist-multi-media-ad-entertainment complex does our thinking for us.

The ultimate object of this complex is to get people to think that events are a result of the "Will of the People." The Will of the People is an engineered reality construct. Who are "the People"? How have they "willed" anything? The end result is all that matters. As Key asserts, once humans believe they think for themselves, "critical postures relax." Once people become uncritical of information being peddled by media outlets, they will accept almost any construction of reality. Americans, says Dr. Key, are "peculiarly vulnerable to such stagecrafted entertainments." Americans, he says, are the most manipulated people in the world.

As Ingo Swann asserts in his book *Penetration,* reality cannot be manipulated without the total and covert management of information pertaining to it. According to Swann, "reality-making" and "information management" are interrelated. He explains that the process of constructing a reality requires the following activities: "the teaching of facts, evidence and information supporting the reality (presented as truth), the disposal of facts, evidence and information which might deconstruct the reality, and the willful introduction of *useful illu - sions* if the above cannot be creatively managed." One who becomes an expert at the above, in modern terms, is called a "spin doctor."

The Information Comfort Zone

Swann asserts that humans have what he terms an "information comfort zone." It is generally observable that most humans do not want information discomfort to be induced. They tend to prefer familiar information which they have processed before, and which is consistent with their agreed-upon ego-based reality, rather than new information which creates discomfort. It has been noted by Aldous Huxley that the human nervous system contains a "reducing valve" of sorts, which protects us from sensory overload. On top of this biological necessity, however, there is a cultural template in operation whereby an overall information comfort zone is established. It would appear that humans maintain an underlying intersubjective "agreement," involving the medium of culture, which involves the "common exclusion of alternate possibilities." We agree to agree on a temporary window on a changing reality; agreement has been called "the cement of social structure."

In order to maintain this social structure, however, the above-described teaching of certain information and disposal of certain information is necessary. The construction and maintenance of a "comfortable" reality requires the creative management of information as well as the "willful" introduction of "useful illusions." The media is just one of many tools used to infuse useful illusions into the public domain, but there are countless creative ways to incapacitate the human mind. Another way to establish an illusory reality structure is to indoctrinate erroneous and outmoded materialist paradigms and limit creative thinking in schools and universities. What we have, essentially, is similar to Big Brother's creation of a completely indoctrinated belief system called "2+2=5," which was set forth so terrifyingly in George Orwell's *1984.*

Perhaps the most startling point Ingo makes, however, is that Earth people seem to be caught up in a "broadly shared amnesia" which, he suspects, is induced in a "wholesale way." This amnesia dulls our memory of events, and keeps us from adding up 2+2. How are we so easily bamboozled? This zombified condition is described in detail by Dr. Key. He writes that by keeping perceptual information isolated, cultural engineers cause people to avoid making certain associations. He points out that journalists, ad-makers, and public relations strategists construct such isolationist strategies so that individuals or target groups will not associate one idea with another. He writes: "isolation can develop out of a cultural system's vested interest in preventing certain conscious connections." Isolationist strategies block what might be considered "reasonable, logical, conscious idea linkages."

The prime pre-condition of subliminal indoctrination by the ad-media corporate alliance is "perceptual rigidity" and "conformity." Once the cultural engineers have the general populace seeing in black and white, cause and effect, simplistic terms, they are primed for the flim-flam man. As Key explains, "one-dimensional, simplistic, cause and effect thinking is one of the most dangerous forces on earth." Physicists have given up looking at the world in a cause and effect way. The idea of causality is now described as an "interconnectedness" between things, with a "non-local" connection or "synchronicity" being more descriptive of reality. It can never be proven that any perceivable effect resulted from a cause; yet, a cause can be searched retroactively until one comes up which fits the desired effect. (As will be discussed in a later chapter, this has also occurred in terms of science's search for human origins and our *surety* that we evolved from the great ape family simply because we can't imagine any other alternative.) As Key writes in *Age of Manipulation*, "North American culture seems to ignore the perceptual reality that much, if not most, of both a cause and an effect will remain unknowable." Cause and effect is a fantasy of the modern Western materialist age.

Swann explains that humans are "information processing entities," and that the average human cannot function very well unless "a few realities are established." I would add that the realities easiest to assimilate are the "factual" and "scientific" ones taught in schools and museums, and paraded before our psyches on the boob tube. A lot of humans do not require "copious amounts of information," Swann adds, and a "fair share of humans will accept illusory information, since it does take rather copious amounts of information to distinguish illusory from other kinds of information." Swann writes: "spin doctors since antiquity have apparently been keenly aware of these two blessings."

Dr. Key's research seems to agree. He asserts that most people have been taught that *decreasing* the number of options is a viable way to arrive at the "truth." This is the equivalent of confining reality within a narrow range of acceptable alternatives. Orwell's *Newspeak* is alive and well as we swerve blindly into the millennium. In addition, as information processors, we are so caught up in the grind of day-to-day survival that we don't have the time it takes to decipher large amounts of illusory information from other kinds of information. Most of us also don't have the intellect to do so, since we are dumbed down by the spin-doctors from birth to death. This is not meant as an attack on anybody's intellect or morality. It takes willful, concerted effort to rise above this heavy weight. To rise to any measure of clarity in this world of externally-based reality constructs is an enormous existential burden.

Swann deduces that the appearance of "possible ET factors" could cause "major wreckage within information comfort zones typical of Earthside information packages." I've deduced that these information packages are ego-driven, and are confined by a heavily peddled materialist scientific paradigm. But who is it, ultimately, that wants our reality confined to this planet? According to Swann, the crux of the matter is that "Earthsiders assume that Earthsiders themselves construct their realities." We assume that it is ourselves, or our social environment, or our media manipulators, but there is an even worse "worse-case" scenario, and Swann is brave enough to suggest it. Swann makes the following rather uncomfortable deduction: "WHO it is that constructs the realities is NOT at all clear." What we have assumed to be an "Earthside" power conglomerate has long known how to indoctrinate the thought processes of humans in order to keep it "earthbound." But, Swann suggests, reality may be engineered *not* by "Earthsiders" but by "Spacesiders," utilizing some type of telepathy or mind control technology.

Interestingly, this subject is alluded to in ancient theories which Immanuel Velikovsky wrote about in 1940. We will get to this after a short discussion of catastrophic theory.

Catastrophic Theory Quick and Dirty

In his book, *Catastrophism, Neocatastrophism and Evolution*, Trevor Palmer outlines the changes which prevailing evolutionary theories have gone through over the past few decades. The Modern Synthesis of neo-Darwinism, which incorporated new genetics developments into traditional Darwinism, and touted gradualist and uniformitarian theories, seemed completely secure in 1959, he writes. Yet, over the ensuing decades, it became clear that the fossil record revealed rather abrupt transitions.

In earlier times, Palmer explains, catastrophism and evolution were mutually exclusive explanations for the fossil record. But after the evidence began to accumulate during the second half of the nineteenth century, he writes, the fossil record could not be explained by the Earth-centered model of catastrophism, which linked such extinctions to global geological upheavals. As a result of distortions propagated by Lyell and others, "generations were led to believe that the views of the catastrophists owed more to preconceived ideas than to observation, whereas the theories of the uniformitarians were all derived by logical deduction from observed data." This myth became widely accepted, Palmer explains, along with the opinion that "catastrophists relied on supernatural explanations for the cause of the major catastrophic events which had supposedly taken place." This assumption derived from the fact that catastrophists generally refused to rule out "the possibility of fresh creations of life at intervals throughout Earth history."

Over a hundred years ago, the idea of the extraterrestrial origin of meteorites was ridiculed. Not until the 1960s was it generally accepted that celestial impacts had caused large craters on Earth. Those who argued that this had anything to do with evolution were derided, even though the fossil record told the story of mass extinction. As Palmer explains: the theory of a supernova explosion was largely ignored, the theory that macromutations might account for the origin of new species provoked derision, and Velikovsky was ridiculed for his theory that the Earth had come into near collisions with planetary-sized bodies. Others established that mass extinctions were real events, but suggested gradualistic explanations. Evolutionary biologists came up with plate tectonics and continental drift to explain them.

The gradualistic view of evolution was challenged in the 1970s by those within the Modern Synthesis who believed that the fossil record showed rapid bursts followed by stasis, or "punctuated equilibrium." With regard to hominid species, however, it has become commonly accepted that "new species appeared in a rapid fashion, and the disappearance of species was also quite abrupt," taking place "against a background of major environmental changes." Today, extraterrestrial impacts are regarded as plausible agents of evolutionary change. Palmer writes that catastrophic events, involving collisions between cosmic bodies, have occurred throughout the history of the Solar System. He writes that the Earth has suffered many impacts, some large enough to cause widespread devastation. He asserts that "ignorance of this extraterrestrial dimension by nineteenth century catastrophists and uniformitarians alike contributed to the demise of catastrophism."

Palmer adds that modern developments in molecular genetics have not revealed any mechanism by which specific genetic changes could be produced in direct response to environmental pressures. He writes that the relationship between evolution at different levels, from molecules to species and above, needs to be clarified, and that the Darwinian Modern Synthesis, "if not actually wrong, is far from complete." The characteristic pattern of evolution is of extinctions followed after a pause by the rapid radiation of new species into vacant ecological space, a very different picture from the one drawn by the founders of the Modern Synthesis. Palmer asserts that, "mass extinctions, whether resulting from catastrophist or gradualistic mechanisms, or a combination of the two, have had a highly significant bearing on the course of evolution."

Yet, nobody is ready to follow the early catastrophists in refusing to rule out the possibility of *fresh creations* of life at intervals throughout Earth history. It is time to once again consider this a possibility.

Velikovsky: How Many Beginnings?

It would appear that certain paradigm-based illusions and constructions have been put in place in order to convince Earth people that there is no exogamous intelligence in the Universe. Darwinian gradualism, as one of the western world's most deeply entrenched materialist paradigms, has contributed much to this illusion.

As discussed earlier, Ingo Swann has suggested that reality may be engineered *not* by "Earthsiders" but by "Spacesiders," utilizing some type of telepathy or mind control technology. The implications of this bizarre circumstance come into better focus once we accept the written accounts of ancient peoples as historical documentation. One of the most important paradigm control issues is the one surrounding the catastrophic history of Earth; a true history involving collisions or near collisions with known bodies in our solar system. These periodic catastrophes resulted in a record of mass extinctions, apparently followed by "macromutations," which Darwin noted was equal to "creation." It would appear that these catastrophes are connected in some way to this "tradition" of paradigm thought control.

"The history of catastrophes is extremely unsettling to historians, evolutionists, geologists, astronomers, and physicists," asserted Immanuel Velikovsky in his unpublished manuscript entitled *In the Beginning*. (www.velikovsky.collision.org) Velikovsky believed that the ancients had witnessed "nature with its elements unchained." This catastrophic past is obscured by the fact that numerous catastrophes have piled up upon each other. Humans enter the theatre, he wrote, at about the third or fourth act. There are, so far, seven acts. Velikovsky reasoned as follows.

The Hebrew cosmogony taught that, before the birth of our Earth, other worlds had been created and destroyed. It is taught that God made several worlds before ours, but he destroyed them all. The Earth underwent reshaping in six consecutive rebuildings, which Velikovsky called "world ages," equaling six major catastrophes. Each of the world ages came to its own doom in a different way. The time in which we live is the seventh creation. It may be interpreted that seven worlds were created, which exist simultaneously and are separated by "abyss, chaos, and waters." However, Velikovsky interpreted this to mean not seven separate worlds in space, but seven kinds of Earth "consecutive in time and built one out of another," forming a unity. He believed that the Hebrew cosmogony was a concept of worlds built and reshaped with the purpose of bringing creation closer to perfection. He suggested that the separation of the seven worlds by "abyss and chaos" referred to the "cataclysms that separated the ages." Velikovsky elucidated this crucial idea when he wrote:

> Even admitting that by 'expulsion from the Garden of Eden' is allegorized a catastrophe which quite destroyed mankind prior to the great deluge, it is impossible to declare that it was the first catastrophe. It depends on the memory of the peoples which catastrophe they consider as the act of creation. Human beings, rising from some catastrophe, bereft of memory of what had happened, regarded themselves as created from the dust of the Earth. All knowledge about ancestors, who they were and in what interstellar space they lived, was wiped away from the memory of the few survivors.

Velikovsky studied in detail the talmudic-rabbinical tradition, which taught that the world was successively inhabited and destroyed even before Adam, the first man, was created. The notion of a succession of worlds created and destroyed is common to many cultures. Max Heindel writes in *The Rosicrucian Philosophy* that the Triune God consists of the Creative Principle, the Preservative Principle, and the Destructive Principle. He writes that forms are created, preserved for a time, and then destroyed so that the construction materials may be used for the building of new forms.

The world ages, each of which lasts a thousand years, are mitigated by deities or personages related to the planets. The idea that the Earth was under the rule of a particular planet at a particular time was taught by various ancient cultures, among them the Pythagoreans, the Magi, Gnostic sects and various secret societies, including the Rosicrucians. Many ancient creeds taught that each of the seven epochs, each lasting a thousand years, is dominated by a different planet and a different personage representing that planet.

Destruction, Creation and the Sabbath

Velikovsky believed that even the naming of the seven days of the week in honor of the seven planets memorializes the seven ages governed by these planets. Many ancient cultures taught that the seven days of the week represent seven world ages, and that each of God's "days" is a millennium. The day of the Sabbath represents the seventh world age. The seventh day, or Sabbath, meant that in six world ages the heavens and the Earth were finally established and, in the seventh age, no further changes in the cosmic order should be expected. The Lord is implored, Velikovsky wrote, to "refrain from further reshaping the Earth." He explained that the word "Sabbath" should be interpreted as *sabbatu*, or "cessation of divine wrath." Velikovsky further explained:

> In the six ages the world and mankind went through the pangs of genesis or creation with its metamorphoses. It is not by mistake that the ages which were brought to their end in the catastrophes of the deluge, of the Confusion of Languages, or of the Overturning of the Plain, are described in the book of Genesis: the time of Genesis or creation was not over until the Sabbath of the Universe arrived. With the end of the world age simultaneous with the end of the Middle Kingdom and the Exodus, the Sabbath of the Universe should have begun.

Velikovsky suggested it was not the deity who provided the example for man to abstain from work on the seventh day, but it is man who "invites the Supreme Being to keep the established order of the heaven and earth, and not to submit them to new revolutions," thus imploring *Him* to take the day off. The "cosmic meaning" of the Sabbath has been lost to mankind, in place of which he inserts a social meaning.

Velikovsky suggested that the identification of major gods with the planets was not symbolic, but, rather, the planets *were* the major gods, and they had a "decisive influence on all physical and moral phenomena of the world." Persian holy books state that "on the planets depends the existence or non-existence of the world." According to the ancient Hebrews, each of seven archangels was associated with a certain planet. The Rosicrucians refer to the planets as the Great Planetary Angels.

Mercury and Memory

Contrary to the common assumption that the planets have occupied the same orbits for billions of years, Velikovsky asserted that the planet Mercury has traveled on its present orbit for only 5,000-6,000 years. He believed that Mercury was once a satellite of Jupiter, or possibly of Saturn, and in the course of Saturn's catastrophic interaction with Jupiter, Mercury was pushed out of its orbit. [Harrington and van Flandern suggested in 1976 that Mercury may be an escaped satellite of Venus.] At some point, Velikovsky suggested, close contact between the magnetospheres of Mercury and Earth occurred, causing a great catastrophe resulting in widescale memory loss in the people of Earth.

In India, it is said that the presence of the god Budha, Mercury, could induce forgetfulness. In addition, the Sumerians believed that it was Enki, or Mercury, who confounded the speech of mankind. The same god, known to the Egyptians as Hermes or Thoth, was known to have "made different the tongue of one country from another." Egyptian scholar, Wallis Budge, has written that Thoth is "speech itself," and that Thoth could "teach a man not only words of power, but also the manner in which to utter them." Even Plato wrote of Thoth, who was the inventor of written language.

Velikovsky outlined the rabbinical conception of seven earths. Each of the seven earths ended with a different type of catastrophe. He wrote that the builders of the Tower of Babel inhabited the fourth earth, which ended in the "sun of the wind." According to rabbinical sources, in the fifth earth, "men become oblivious of their origin and home." Simultaneous with the speaking of a new language, the tower builders forgot their own language and forgot their own past. This generation is called "the people who lost their memory." The Earth they inhabit is "the fifth earth, that of oblivion." Mexican sources also say that those who survived the catastrophe of the wind "lost their reason and speech."

Velikovsky explained that the worldwide characteristic of the catastrophe of the "sun of wind" was "its influence upon the mental, or mnemonic, capacity of the peoples." He wondered if the Earth underwent some type of electromagnetic disturbance, and whether the experience was the equivalent of a deep electrical shock which affected certain areas of the brain. According to Ginzberg, in *Legends*, the place where the Tower of Babel once stood still induces total memory loss in anyone who passes by.

Velikovsky believed that the planet responsible for memory loss disturbances is the planet Mercury, affiliated with the Greek God Hermes. Mercury is also known by the Sumerians as Enki, by the Egyptians as Thoth, and to the northern peoples as Odin. In many astronomical texts, Mercury is assigned the dominion of memory and speech in mankind. The Romans and Greeks pictured Mercury with wings, as well as with an emblem, the caduceus, a staff with a double serpent (as used by both the Serpent Staff Pleiadians and the American Medical Association). Mercury was also known as the messenger of the gods.

The Tower of Babel and Alternative Three

Velikovsky's writings bring to our attention another piece of the puzzle. The Tower of Babel story existed in the remotest parts of the world prior to missionary visits to these areas. In his manuscript, *In the Beginning*, (www.velikovsky.collision.org) Velikovsky outlined the catastrophe of Babel and the subsequent confusion of languages. According to rabbinical sources, he wrote, the high tower was built as a shelter in case another great deluge should occur. The Babylonian account explains that when it was "close to heaven," the gods "sent winds and ruined the entire scheme." The Mexican version of this catastrophe verifies this assumption. In order to protect themselves from this epoch world destruction, a very high tower was built. At a crucial moment, their languages were changed, and they couldn't understand each other. There was also a crack, like a thunderbolt, from above. From then on, a pervasive confusion and amnesia set in. The people were dispersed to many lands where, to this day, their beliefs and ways are worlds apart and communication between them is difficult.

The memory of various cataclysms caused mankind to be concerned about the future, and seeing that the gods survived by moving themselves off the planet, they attempted to put their fate in their own hands. The gods of Sumer couldn't have this. The Tower of Babel was not simply a "tower;" it was a pyramid. Alternative opinions regarding the great pyramids assert that they were huge shelters for the "elite." Others assert the flat surfaces on the top were landing pads for the flying machines of the gods. The Tower of Babel incident, then, may be seen as mankind in the act of mimicking the gods *in very distinct ways* involving survival and, possibly, space travel. Modern man's ventures into space may be doomed in the same way.

The International Space Station seems to be a huge outer space replica of the Tower of Babel, wherein we once again attempt to expand our existence beyond our blue oasis. The Babel story indicates that, in the historical past, human attempts to move off their sphere for the safety and continuation of the race were dashed upon the rocks. From this point of view, the Tower of Babel incident is mindful of an "Alternative 3" type of scenario. By mimicking as best they could what the gods did during the Great Flood, humans were not merely attempting to gain some type of vague, lofty position, but were actually trying to avoid their own decimation by lifting themselves off the planet. The International Space Station is an effort to do the very same thing.

According to NASA's Astrobiology Roadmap, Objective 16 is to "understand the human-directed processes by which life can migrate from one world to another." Objective 16 is also clear in stating that the results of the engineering of artificial ecosystems in space will answer fundamental questions about whether life is purely a planetary phenomenon or whether life is able to expand its evolutionary trajectory beyond its home planet. In this regard, the near to mid-term implementation of this objective is to "use low Earth orbit opportunities as a testbed for studying evolution and ecological interactions in the space environment." This is to include "wild biota indigenous to the spacecraft," and will attempt to determine how to promote evolutionary success, and how to extend these opportunities to other planetary bodies in concert with human exploration of the solar system.

It is clear that Objective 16 will have all these opportunities and more within the International Space Station. As discussed earlier, NASA's Astrobiology objectives include the following: "engineer closed and open environments as prototypes for human exploration of other planets, test such systems in analog environments on Earth and in space, place candidate ecosystems on extraterrestrial surfaces and document their evolution, establish permanent colonies of humans and other organisms in space and on another planetary surface, engineer life for survival, adaptation and evolution beyond Earth."

As we will soon see, there is a curious tie-in between NASA's attempts to move earth life to another planet, and the seemingly religious and ritualistic manner of achieving these objectives. Are they seeking permission, or have they already acquired permission, for liftoff to Galactic humanhood?

The Mercury Seven

As Mary Zornio writes in *The 40th Anniversary of Mercury 7* (www.hq.nasa.gov/office/pao/History) Virgil I. Grissom was in the U.S. manned space program since 1959. Following his Air Force service in Korea, Grissom became one of NASA's Original Seven Mercury Astronauts. The Mercury Seven—Gus Grissom, Scott Carpenter, Gordon Cooper, John Glenn, Walter Schirra, Alan Shepard, and Donald Slayton—became instant heroes after they were introduced to the public on April 9, 1959.

Grissom was the second U.S. citizen in space, and the first man to fly in space twice. His first space flight was for Mercury 4 on July 21, 1961 in his own capsule, the *Liberty Bell 7*, which he named for its bell shape. On his second space flight, on March 23, 1965, he served as commander for Gemini III, the first manned Gemini flight. Grissom had been told privately that if all went well, he would be the first American to walk on the moon. He wrote in his 1968 book, *Gemini: A Personal Account of Man's Venture Into Space*, that he was very fortunate to be participating in such a "weird, wonderful enterprise."

Zornio explains that Grissom and his colleagues worked long hours, and kept their eyes on the prize: to be the first nation in space. However, that prize was snatched out from under them on April 12, 1961, when Russian cosmonaut Yuri Gagarin completed a one and three quarter hour orbital flight. On May 5, 1961, Alan Shepard became the first American in space when he successfully piloted a fifteen minute suborbital flight on board the *Freedom 7*. Grissom was selected to fly the second flight. As M. Scott Carpenter wrote in *We Seven*, Grissom named his MR-4 spacecraft *Liberty Bell 7* "because the capsule does resemble a bell."

As Carpenter wrote, the *Liberty Bell* had three significant improvements over Shepard's spacecraft. The control panel had been redesigned, a large picture window had been installed, and the *Liberty Bell* had an innovative explosive hatch. The hatch was held in place by seventy bolts and was opened by triggering a Mild Detonating Fuse, which delivered a blow to a special plunger. Once the pilot activated the mechanism, the hatch would blow off the spacecraft so that the pilot could get out of the capsule quickly. However, the new hatch had not been previously tested.

The *Liberty Bell* was given the go ahead for launch on July 21, 1961. Everything went well with the fifteen minute flight, except that, as reported in *We Seven*, Grissom described the attitude controls as "sticky and sluggish." The G forces in the

The original Mercury Seven astronauts in an archive photo. NASA Photo Archives.

capsule reached a peak of 11.2 during re-entry, but were not a major problem for Grissom. After splashdown, Grissom began preparations to exit the craft. As reported in *We Seven*, Grissom began to detach himself from his harnesses, straps, and medical sensors, and then began standard procedures for arming the detonator which would open the hatch.

Zornio explains that Grissom then notified the recovery helicopter that he would need a few minutes to mark all of the switch positions on the capsule's instrument panel. He then

Gus Grissom, one of the original Mercury Seven Astronauts, the second man in space, and the first man to go into space twice, was expected to be the first man on the moon.

informed the helicopter that he was ready. As he later stated, "according to the plan, the pilot was to inform me as soon as he had lifted me up a bit so that the capsule would not ship water when the hatch blew. Then I would remove my helmet, blow the hatch and get out."

Grissom waited for final confirmation to blow the hatch, when suddenly, as reported in *We Seven*, "the hatch blew off with a dull thud," and water began to flood the capsule. Grissom threw off his helmet, and quickly pulled himself out of the sinking craft. Once in the water, Grissom became tangled up in the lines, and for several panicked minutes he remained attached to the sinking *Liberty Bell*. He was finally able to free himself from the lines and swim away from the capsule. But he still wasn't out of trouble.

As Grissom's suit quickly lost buoyancy due to a leak, he floundered in the water. Unaware of the difficulty Grissom was having in the water, and consumed with their efforts to save the *Liberty Bell*, Grissom later reported, neither of the two helicopters seemed aware that he was urgently in need of a life line. As exhaustion set in, Grissom thought to himself, as reported in *We Seven*, "well, you've gone through the whole flight, and now you're going to sink right here in front of all these people." Filled with water, the *Liberty Bell* was too heavy to be lifted by helicopter. The extra weight put too much strain on the helicopter and a red light warned that engine failure was imminent. The recovery team was forced to let the *Liberty Bell* go.

Finally, a third helicopter dropped Grissom a line. Aboard the carrier, Grissom received a telephone call from President Kennedy expressing relief that he was safe. However, Grissom was very disappointed that he had come back without his spacecraft. As reported in *We Seven*, Grissom claimed, "It was especially hard for me, as a professional pilot. In all of my years of flying— including combat in Korea—this was the first time that my aircraft and I had not come back together. In my entire career as a pilot, *Liberty Bell* was the first thing I had ever lost."

The press conference which followed the *Liberty Bell* flight turned out to be an uncomfortable experience because, as Grissom's wife Betty later wrote in *Starfall*, "the reporters skipped over the successful aspects of the flight... and probed around the question of whether Grissom had contributed to the loss of the *Liberty Bell* by accidentally bumping the plunger which blew the hatch." As

66

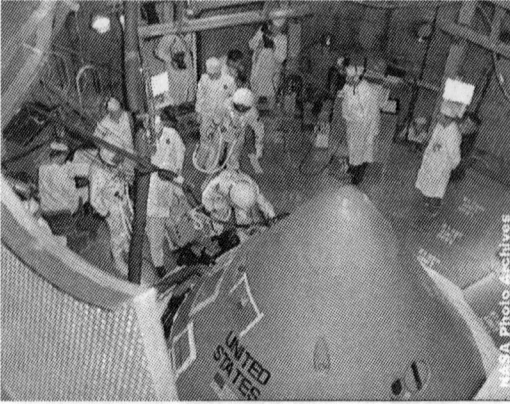

The Apollo 1 astronauts, Gus Grissom, Ed White, and Roger Chaffee as they enter the capsule before the fateful fire. NASA Photo Archives.

reported in *Moon Shot* (Turner Broadcasting, 1994) Grissom explained: "I was just laying there minding my own business when, POW, the hatch went. And I looked up and saw nothing but blue sky and water starting to come in over the sill."

Although a review board determined that Gus did not contribute in any way to the premature detonation of the hatch, questions surrounding the incident refused to go away. According to Betty Grissom in *Starfall*, "engineers spoke of a transient malfunction but were helpless to identify it because the capsule and the hatch were now on the bottom of the ocean." Grissom is quoted in *We Seven* as saying:

We tried for weeks afterwards to find out what had happened and how it had happened. I even crawled into capsules and tried to duplicate all of my movements, to see if I could make the whole thing happen again. It was impossible. The plunger that detonates the bolts is so far out of the way that I would have had to reach for it on purpose to hit it, and this I did not do. Even when I thrashed about with my elbows, I could not bump against it accidentally.

Grissom was then assigned to the Gemini program, with Alan Shepard as commander. It was said of Grissom that he combined his skills in mechanical engineering and test piloting to help produce a manned system designed to rely on the input of the pilots. As Don Slayton wrote in *Deke! US Manned Space From Mercury to the Shuttle*, "Gemini would not fly without a guy at the controls... It was laid out the way a pilot likes to have the thing laid out... Gus was the guy who did all that."

Curiously, commander Alan Shepard began to experience severe nausea, vomiting and dizzy spells, which was diagnosed as an inner ear disorder called Meniere's Syndrome. Grissom took over the commander's seat, and the pilot's seat went to John Young. Grissom named the Gemini craft *The Unsinkable Molly Brown*. John Young was later reprimanded for stealing on board the *Molly Brown* a corned beef sandwich, which he shared with his captain. The *Molly Brown* flew 80,000 miles and completed three successful orbits around the earth.

In March 1966, NASA introduced Gus Grissom as commander of the first Apollo Earth-orbit mission. By the time Gus was freed up from his Gemini duties, the Apollo spacecraft and its systems were ready for testing. Unlike the full

The Apollo 1 space capsule after the fire that killed three astronauts on **January 27, 1967. NASA Photo Archives.**

control he was used to, however, Grissom and his crew "inherited a spacecraft that had been designed for them, but not with them." Grissom, along with crew members Ed White, Roger Chaffee, and their support engineers and technicians, were only able to participate in the inspections of the craft. As Betty Grissom wrote in *Starfall*, "some of the things Gus saw he did not like."

Grissom's wife reported in her book that Gus was very troubled about the spacecraft. He became concerned that several engineering changes were incomplete. For instance, she reported, the environmental control unit "leaked like a sieve," and had to be removed, and the training simulator was not in any better shape. As astronaut Walter Cunningham stated in *Moon Shot*, the spacecraft was "in poor shape." Cunningham asserted, "it just wasn't as good as it should have been for the job of flying the first manned Apollo mission."

In a brief stop at home before returning to the Cape, Betty Grissom wrote, Gus pulled a large lemon from a tree in his backyard, and, as he kissed his wife goodbye, explained: "I'm going to hang it on that spacecraft." Little did she know, the craft truly was a lemon, and she would never see her husband again. On January 27, 1967, Gus Grissom, Ed White and Roger Chaffee died in a suspicious fire aboard the Apollo 1 space craft.

NASA's Symbolic Rituals

Richard Hoagland and Michael Bara of The Enterprise Mission have been gathering evidence of what they call "the true history of NASA and its hierarchy." They have traced in detail the celestial alignments during various NASA space missions, and they have discovered what they believe to be "constant offerings to the gods of ancient Egypt." Hoagland and Bara have come to the conclusion that space missions have some kind of "agenda," which seems to act out a symbolic religious ritual.

Hoagland and Bara refer to the Biblical story of Armageddon, the final war which will devastate the planet, and after which Christ will return to Earth for his reign of a thousand years. This story, they assert, is the same story as the Egyptian planetary gods Osiris and Set, wherein Horus, the son of the god Osiris, comes to Earth to do battle with Set, the embodiment of evil. A war will decide the path of mankind and restore Osiris to his earthly kingdom. The point in either case, they assert, is that the final war needs to be *started* in order to bring about the "spiritual transformation" of mankind.

Nearly 38 years to the day that it was lost at sea, Liberty Bell 7, the capsule of astronaut Virgil I. "Gus" Grissom, was raised from the floor of the Atlantic Ocean. The Liberty Bell was lost on July 21st, 1961 when the hatch prematurely blew off after splashdown. The spacecraft had remained at the bottom of the Atlantic until a salvage expedition was launched on April 20th, 1999. This date is one of the repetitive "ritual dates" which, Hoagland and Bara have noted, comes up repeatedly in NASA history. Other July 20 dates being celebrated that week were the July 20, 1969 Apollo 11 Moon landing, and the first successful landing on Mars on July 20, 1976. (see "Table of Coincidence" at their web site)

Grissom was one of the original Mercury 7, and was also a 33° Scottish Rite Freemason. Despite the accident, Grissom was slated to be the first man on the Moon. As commander of Apollo 1, he was later tragically killed in the fire which also killed astronauts Roger Chaffee and Ed White on January 27, 1967. On February 11, 1999, the AP reported that Grissom's son, Scott Grissom, had alleged, "My father's death was no accident. He was murdered." The press release also explained that lead NASA investigator, Clark MacDonald of McDonnell-Douglas, has also charged that the agency engaged in a cover-up of the true cause of the fire by destroying his original report and interview tapes of thirty-one years ago. Grissom's son believes the motive for his father's killing was related to NASA's desire not to allow his father to be the first to walk on the Moon.

Richard Hoagland had earlier suggested that the fire that killed the astronauts had been a strange coincidence, in that the fire was seemingly coordinated to take place at the same moment as a White House reception honoring the signing of the UN Space Treaty of 1967.

In his book, *Lost Moon*, commander of Apollo 13, Jim Lovell, had also noted this odd coincidence. He wondered why: "The invitation said 6:14 p.m. It actually said that. Not 6:15, or 6:13, but 6:14. It was unclear to me what they were going to do with the extra minute."

Hoagland and Bara assert that the fire deliberately occurred precisely in the middle of the signing ceremony. As the fire roared through the command module, they write, "the doomed astronauts tried vainly to get out, but the stars had already sealed their fate." As the ceremony was taking place, they write, "in the skies above Washington, Alnitak in Orion's belt was at 33°." In addition, they note, the view from the cape where the fire broke out, "revealed Comet Encke, the lead object in the Taurid Meteor stream, at 33°." And, at the "then future Apollo 11 landing site, where Grissom would have taken those first steps into immortality, Alnitak in Orion's belt was dead on the horizon." And the view from Giza, "the linch pin of the entire NASA/Egypt/Masonic connection, found Alnilam at the magical tetrahedral location of 19.5°." Hoagland and Bara assert:

> All of this is richly symbolic and consistent with our previous decoding of NASA's celestial hi-jinks. To have Encke, the veritable embodiment of chaos and death in the Egyptian system, hanging over the Cape as the fire broke out is as strong an omen as one could find. The symbolic message of all this seems to be "AMason is about to be sacrificed." But why was he sacrificed, and why is his memory now being "resurrected" on a ritual date?

Hoagland and Bara conclude that someone in the Johnson White House was not interested in having one of Kennedy's men be the first to set foot on the Moon. Apparently, they surmise, someone was willing to end the lives of three people to obtain this goal.

The Symbolic "Raising" of Liberty Bell 7

As the rest of the nation focused on the fatal plunge of JFK, Jr.'s plane into the waters of the Atlantic on July 16, 1999, and his subsequent quick and secretive burial at sea on July 21, 1999, along with his two compadres (another trio of innocents), Liberty Bell 7 was raised from the ocean bed at 2:15 a.m. on July 20, 1999. Hoagland and Bara write that the recovery "had the smell of a ritual procedure written all over it right from the beginning." They were surprised to find even more correlations as they moved through the solar system "temples" set up by NASA.

> Finding Orion's belt star Mintaka at 19.5° below the horizon when viewed from the recovery site was not at all unexpected. ... At the same moment at Giza, Regulus, the "heart of the lion" and representative of the Egyptian god Horus, was 19.5° over the plateau ... and on Mars, at the Viking 2 landing site, we found Sirius (Isis) at 33° below the horizon. We also found yet another interesting alignment at Cydonia itself, with the comet Encke, the lead object in the Taurid meteor stream, at -19.5°.

> In essence, the "Resurrection" of Grissom's capsule took place at a specific time and place so that all of the key gods in the ancient Egyptian pantheon could observe from "sacred positions" in the heavens. This overwhelmingly symbolic act was played out before Osiris, the god of resurrection, at the site of the recovery. Horus, the son of Osiris was at Giza, Isis, mother of Horus and wife/sister of Osiris was at the Viking 2 site, and Set, as represented by Taurus/Encke was at Cydonia. In the years that we have been tracking this

NASA/Egypt connection, rarely if ever have we seen such a complete roll call of the Gods of the Nile.

This pantheon of celestial gods, they explain, Isis, Osiris, Set, and Horus, are all part of the same story, stemming from the mysterious "Zep-Tepi" (The First Time). Strangely, at least nine (1942) Mercury dimes were found in the capsule. On this dime, Mercury (a.k.a. Thoth/Hermes), the "winged messenger," is pictured. According to Velikovsky, following the gift of the written word to mankind by the god Mercury, mankind lost the capacity for the oral transmission of knowledge. This resulted in widescale memory loss, in particular, with respect to understanding our ancient connection to the Universe. I believe this should be understood in the context of a general oblivion and widescale ignorance of "occult" or arcane knowledge, specifically with regard to human origins, and as well, the connection between human origins and the planets, the personages represented by the planets, and the "fate" of humankind.

How accurately this mythological countenance describes the media-induced pall that strained the collective psyche during this strange week. Disconcerting events and images were coming at us a mile a minute. There was a sense that a ritual of some kind was taking place, but we were unable to make the conscious idea linkages necessary to verbalize it.

Just days before JFK, Jr.'s death, it was reported in the *Washington Post* that JFK, Sr.'s original casket had been dropped into the Atlantic Ocean from an airplane a few years following the assassination. We will probably never know what was in that casket, but some wonder if it was real body of the slain president. The article reported that assassination researchers were incredulous that this item of obvious importance to the case had been discarded in such a bizarre manner without public discussion. Following this story was the report that JFK, Jr.'s plane was missing, and the strange ritual removal of three bodies from the water only to return them again.

John F. Kennedy, Jr. speaks with members of the national media at the HBO/Imagine Entertainment premiere of "From the Earth to the Moon" at the John F. Kennedy Space Center on March 23, 1998.

Why were three civilians given an immediate burial at sea? Other auspicious dates being celebrated that week were the July 20, 1969 Apollo 11 Moon landing and the first successful landing on Mars on July 20, 1976. In addition, that week was also the anniversary of the Chappaquiddick incident, an accident involving another Kennedy plunge into the same waters of the Atlantic, and a woman's life offered up to the gods.

In an anonymous e-mail which I received that week, the writer noted the following:

In one of his rare references to a specific calendar date, Nostradamus predicts that in the seventh month of 1999, a man described as the "King of Terror" will descend from the sky: 'The year 1999, seventh month, A great King of terror will come from the sky: To resurrect the great King of Angoumois, Mars to reign fortunately before and after.'—Century X, Quatrain 72

Angoumois is a French region (just north of Bordeaux) whose principal city is Angouleme. Angouleme gave its name to the dynasty of kings who ruled

France during Nostradamus' lifetime. In the context of the prophet's times, therefore, "Angoumois" would describe someone belonging to the royal family. In a related Quatrain, Nostradamus envisions a lamentable event which causes the cancellation of a wedding involving a first cousin of this royal family:

'Lamentable cries will come then from Angouleme, And the marriage of the first cousin will be precluded.'—Century X, Quatrain 17

Could these visions be referring to the disappearance of John F. Kennedy, Jr., the most charismatic scion of America's "royal family"? Consider that: (a) He literally "fell from the sky" on July 16, 1999; (b) Before and after his disappearance, Mars was passing through Scorpio, one of the two astrological signs in which that planet "reigns"; and (c) He disappeared en route to the marriage of his first cousin, which was postponed in reaction to the tragedy.

If JFK, Jr. is Nostradamus'"King of Angoumois", then we must anticipate (per X.72) that he will be "resurrected." This may signify that he will somehow be discovered alive after being presumed dead. The following verses seem to say as much:

'... the Prince with a stump foot, fear of enemies will make his sail bound.'—Century III, Quatrain 91

The image of the 'stump foot'evokes the cast which JFK, Jr. was wearing on his right foot when his plane crashed. Does "fear of enemies" suggest that his flight was sabotaged? And did he perhaps rig a crude sail to escape from the wreckage scene? Nostradamus also drops some clues as to when JFK, Jr. may be expected to reappear. In the Quatrain which follows the one quoted above, he says that the planet Saturn will come back again late. Oddly, the 'resurrection' which Nostradamus foresees for JFK, Jr. does fit the scenario laid out in Chapter 13 of the Book of Revelation, where the fatally wounded Beast emerges from the sea and is miraculously revived.

As hindsight is 20/20, we can see that it was not JFK, Jr. who was miraculously revived, but it was the fatally wounded Liberty Bell 7 which emerged from the sea only a few days later and is currently being miraculously revived. In keeping with this "predilection for ritual and reverence," and to reinforce the cyclical nature of the symbolic ritual, write Hoagland and Bara, the Liberty Bell was returned to Port Canaveral on July 21, 1999, exactly 38 years to the day that it had left. (Interestingly, JFK, Jr. was 38 years old when he died.) They write:

> The position of Encke at Cydonia is especially interesting. As we head into the November passage through the Taurid Meteor Stream, why the sudden need to "resurrect" the fallen Masonic hero on the same weekend that Kennedy's son [the son of the only Catholic president in U.S. history] was lost in an apparent accident? What cycle is about to be completed?

The International Space Station

Although NASA and other space agencies and astronomers normally have an obvious predilection for naming things, a name for the "International Space Station" has not been forthcoming; at least not until it was unofficially christened by Hoagland and Bara: "IS(I)S."

On November 20th, 1998 the first module of the new International Space Station (IS(I)S) was launched at 6:40 AM GMT from the Baikonour Cosmodrome in Kazakstan. The timing of this launch, Hoagland and Bara assert, was "evidently designed to coincide with a number of significant celestial alignments consistent with NASA's long established ritual pattern." The precise meaning of these alignments, they explain, take on different characteristics depending on the place in which you are standing.

The Russian module, Zarya, which translates into "sunrise" or "rising sun" (Horus) in English, was launched from pad 333 at its precisely scheduled time despite Russian requests to have the launch delayed. NASA refused the request

Phase 1 of the planned International Space Station. Illustration commissioned by Johnson Space Center in 1993. NASA Photo Archives.

and the launch went off in November of 1998 as originally scheduled. Hoagland and Bara assert that another minor delay would not have presented a huge problem, but the symbolic significance of the date seems auspicious. They write:

> 'Rising sun' of course, is Horus in the ancient Egyptian pantheon of gods. "Horus in the Horizon" or more literally "Horus-rising," is symbolic of the transformation between the dimensions of life and death. To the Egyptians, the symbol of their sun-god rising each morning was a comforting reminder that the good god Osiris ruled over the underworld to the west and that rebirth was promised. Recent studies have connected Horus directly to the Sphinx and to the planet Mars, which shared the name Horatkhi, which means "Horus in the Horizon." The Sphinx was also painted red for much of its documented history, and this also ties quite nicely with the planet Mars which was known as "Horus the Red" to the Egyptians.

A short time later, Horus (son of Isis and Osiris) docked with "Unity" in December of 1998, thus "unifying" Isis and Horus.

The number 33 is the top level achievable in Scottish Rite Freemasonry. Many executives in NASA, as well as NASAastronauts, are Scottish Rite members. The number 33 and its variants also have an unusual significance in the space program, and seems to "pop up everywhere," such as:

- The landing site of Mars Pathfinder on "Horus the Red" is located at 19.5° N by 33° W.

- The launch pad at the White Sands Missile range where Werner Von Braun ran the post war V-2 tests was launch pad 33.

- The landing strip at Kennedy Space Center in Florida where the shuttle lands is called "Runway 33."

- The symbolism of "Horus" launching from pad 333 (which just happens to be 3,300 km from Giza) is fairly obvious.

The 6:40 AM GMT launch time was a moment of incredible celestial convergence over Baikonour. One source gives the location of the Cosmodrome as 45° 56' N by 63° 18' E. This might be considered the center of Baikonour, but the locations of pads and other facilities are spread out. Assuming this latitude and longitude to be correct, Hoagland and Bara note that "Sirius/Isis, consort of Osiris and mother of Horus, namesake of the International Space Station, was at precisely 33° below the western horizon.

Hoagland and Bara note other alignments taking place above the launch site at the Cosmodrome. They conclude that the recurrent astronomical theme is that of Mars. They maintain that NASA's reason for selecting their specific launch date and time was to coincide with certain celestial events. Hoagland and Bara insist that there is a message in this "signal," and that the "signal is *increasing* in both *strength and frequency,* perhaps implying that we are near the brink of an event of major significance." It may be "the joining of Zarya and Unity to form IS(I)S," or it may be something more significant. They note that one of the models for a manned mission to Mars was to strap engines on the IS(I)S and go there. Is this the ultimate aim of the Mars references?

Hoagland and Bara assert that not only was Sirius at 33° 33', symbolic of Freemasonry's highest level, but they emphasize the fact that it *was* Sirius, the celestial counterpart of the Goddess Isis, rather than another one of the "NASA deities." They wonder, could this be translated as: "a Mission *to* IS(I)S *with* Isis at 33° 33'?" Following the same pattern with this mission, they noted that, from the ground when the shuttle was directly overhead, the gods were once again in their symbolically appropriate positions. According to their analysis, this celestial entourage represents "Orion/Osiris Rising" and giving birth to new life, as Isis, the "sister/wife/consort" and "Mother of the Nile" passes overhead. They write: "We are by now used to this sort of thing from NASA. This kind of symbolism and ritual are imbedded deeply into the agency's every move."

The Mission Patch

Hoagland and Bara write that the Mission Patch for STS-96 is overwhelmingly symbolic. The Egyptian Goddess Isis is the same as the star Sirius. Isis, both the personage and the star, were

> The STS-96 Mission Patch can be viewed at www.enterprisemission.com and purchased online at www.astronomynowstore.com. All NASA mission emblems hold a copyright by NASA, but can be seen on line in the NASA photo journal archives.

represented by several hieroglyphs, including the cobra, or serpent, and the serpent's tooth. Hoagland and Bara write: The tooth is a vertically elongated triangle and the cobra is coiled as if to attack. Taken together, they literally translate to 'the Goddess Sirius.'" The Mission Patch shows the same motif, a vertically elongated triangle with a red, white and blue elongated triangle pointing above to a single star. Could this star represent Sirius? Hoagland and Bara explain that the five pointed star, or pentagram, and the triangle pointing to the star is encircled by an orbital trajectory that is elliptical. This elliptical orbit is more representative of a planetary or stellar orbit, they suggest, rather than a circular satellite orbit. They wonder if these symbolic links allude to the work of Zecharia Sitchin.

Hoagland and Bara clarify that IS(I)S is encoded three separate ways on the patch: as a triangle, a star and as the space station itself. They explain: on the patch is indicated the "Planet of the Crossing," Sitchin's mythical 12th planet, and, as well, an elliptical orbit, which is a further reference to the 12th planet. Another image on the patch shows the Earth tilted off its proper axis. This image of the Earth is a reference to the pole shift, and the suggested role of the 12th planet in causing the pole shift that instigated the Great Flood. Hoagland and Bara believe the Mission Patch indicates that the real purpose of the International Space Station is to search for the 12th planet. The space station will also become the means to effectuate indefinite survival in the case of a planetary catastrophe.

The authors conclude: "Symbolically giving birth to new life while await-ing imminent disaster is a deep and recurring theme running through our plan-et's mythos. NASA seems to be building on this same theme."

In light of the separation of church and state, why is a civilian agency prac-ticing ritual magic before the eyes of an uninitiated secular public, a public which has no context for understanding arcane connections of this nature? As Michael Hoffman has written in his now infamous treatise *Secret Societies and Psychological Warfare*, the "cryptocracy" is in the alchemical process of what he calls "Revelation of the Method." He asserts that exposing the cryptocracy without a broad understanding of the "hermetic-alchemical" control process is useless.

As "Diogenes" asks in "The Only Way Out is Through the Hidden Door," (in *Paranoid Women Collect Their Thoughts*):

> Why are these men exposing themselves? What kind of conspiracy is this, that can't even keep its own secrets? ... It's one so certain of its perverse preemi-nence that it's rubbing our noses in its private parts. ...

As David Icke, author of *And the Truth Shall Set You Free*, and *The Biggest Secret*, has suggested, this behavior is connected to that mysterious ancient group known as The Illuminati. In his essay entitled "A Concise Description of the Illuminati" (www.davidicke.com), Icke describes in detail this powerful clique which controls events on the 3rd dimensional plane from the lower 4th plane. Icke believes that these 4th dimensional entities are reptilian, and that they work through persons in this dimension who have a vibrational compati-bility. This compatibility is essentially a blood relation. He indicates that this blood lineage runs through European royal and aristocratic families as well as Eastern Establishment families of the U.S. As Icke writes:

> Every presidential election since and including George Washington in 1789 has been won by the candidate with the most European royal genes. Of the 42 pres-idents to Bill Clinton, 33 have been related to two people, Alfred the Great, King of England, and Charlemagne, the most famous monarch of what we now call France. It is the same wherever you look in the positions of power - they are the same tribe.

In addition to this obsession with blood lineage, Icke writes, members of the Illuminati also have an obsession with symbolic ritual. He even points out a cor-relation between what science has noted about the "reptilian" part of the human brain (the "R-complex") and the behavior traits of the Illuminati mentality: an obsession with ritual, cold-blooded behavior, territorialism, and an obsession with top-down hierarchical structures. He explains what the secret Illuminati rit-uals aim to achieve:

> Their ritual is not just for ceremonial purposes or gratuitous horror. The rituals are designed to rewire the energy fields and grids of the planets and therefore to fundamentally affect human consciousness. The rituals these bloodlines per-formed in the ancient world are the same as they do now. They have a detailed annual calendar of events on which they perform their sacrifice rituals in line with key lunar, solar, and planetary cycles to harness that energy for their sick agenda to take complete control of Planet Earth in the very near future.

As Eugenia Macer-Story has discovered in her vast research into discarnate intelligence, "human beings regularly experience the intrusion of powerful interdimensional 'spirit' entities with power appetites which can be puzzling." In her essay entitled "The Dark Frontier: Master Game Plan of Discarnate Intelligences" (in *Paranoid Women Collect Their Thoughts*), she notes that neg-ative intelligences "will often be found in power positions which tend to under-mine the functioning of democratic or constitutional systems of government which rotate power positions among a variety of individuals." As Icke has suggested, the way to keep this power in the same lineage is to keep the blood line intact. That secret blood line is called The Illuminati. Macer-Story's extraordinary research seems to back up the idea that discarnate intelligences do in fact tend to aim for hereditary power. She writes:

> If, in a family or small business situation, a power-oriented discarnate intelligence can enter telepathically the minds of the persons in authority, this intelligence can 'rule'the limited group of blood relatives and/or employees with a dictatorial energy ... This manipulative discarnate intelligence can then 'use' the family as a material 'chess piece' with which to enter the larger material arena in order to obtain dominating power over events. An executive of a larger firm or a member of a family with hereditary aristocratic and/or economic status, when dominated by a power oriented discarnate intelligence, constitutes a 'Knight' or 'Bishop,' or 'King' or 'Queen' on the material chess board. Symptoms of takeover by negative discarnate intelligence should therefore be watched for within political situations involving monarchy, dictatorship and/or significant hereditary financial power...

> One motivation for dominating a series of perishable organic bodies might simply be the desire for continuation of this power-oriented identity in sequential time. Like the Pony Express which employed riders who galloped one horse to exhaustion, then leapt onto a fresh horse to continue the route, inter-dimensional intelligences may dominate a fresh and suitable new organic host in order to continue on the same pre-intended historical course.

What might all of this have to do with NASA and Freemasonry? The connection lies in building edifices on top of sacred power points, and in the ritual control of the powerful energies released at these sites by such ritual "building" behaviors. To imagine this, we must keep in mind that NASA's Astrobiology tenets aim to construct human habitats which will take us off planet. In this instance, we may not be building the sorts of constructions that were built in ancient times, but, as in the case of the International Space Station, we are building a home away from home. We are building the Tower of Babel. To get the gist of the connection to ancient Mayan sacred energy traditions, consider the following quote from a website called Sacred Sites (www.sacredsites.com):

> To the Maya, the world was alive and imbued with a sacredness that was especially concentrated at special points, like caves and mountains. The principal pattern of power points had been established by the gods when the cosmos was created. Within this matrix of sacred landscape, human beings built communities that both merged with the god-generated patterns and created a second human-made matrix of power points. The two systems were perceived to be complementary, not separate. ....

> The world of human beings was connected to the Otherworld along the wacah chan axis which ran through the center of existence. This axis was not located in any one earthly place, but could be materialized through ritual at any point in the natural and human-made landscape. Most important, it was materialized in the person of the king, who brought it into existence as he stood enthralled in ecstatic visions atop his pyramid-mountain....

> When new buildings were to be constructed, the Maya performed elaborate rituals both to terminate the old structure and contain its accumulated energy. The new structure was then built atop the old and, when it was ready for use, they conducted elaborate dedication rituals to bring it alive....So powerful were the effects of these rituals that the objects, people, buildings, and places in the landscape in which the supernatural materialized accumulated energy and became more sacred with repeated use. Thus, as kings built and rebuilt temples on the same spot over centuries, the sanctums within them became ever more sacred.

> The devotion and ecstasy of successive divine kings sacrificing within those sanctums rendered the membrane between this world and the Otherworld ever more thin and pliable. The ancestors and the gods passed through such portals

into the living monarch with increasing facility. To enhance this effect, generations of kings replicated the iconography and sculptural programs of early buildings through successive temples built over the same nexus....As the Maya exploited the patterns of power in time and space, they used ritual to control the dangerous and powerful energies they released.

NASA's Nazi War Criminals

To imagine that someone at NASA may possibly be orchestrating a pattern of ritualistic sacred "gematria" borders on insanity if we are ignorant of the hermetic-alchemical control process. Magic is but misapprehended Science. We must consider that most of NASA's personnel may not even be aware of these ritualistic patterns. As David Icke's information might indicate, the timing of these incidents could very well be controlled from an invisible "platform," located in the lower fourth astral plane. And, as Eugenia Macer-Story's information might suggest, such a pattern of negative influence can be exerted over time through successive or generational influences in a family or large company.

The American space program owes much of its success to the recruitment of Nazi war criminals after World War II. According to Lee and Schlain in *Acid Dreams*, the U.S. Naval Technical Mission brought home every piece of scientific data that could be salvaged from the fallen Reich, including detailed reports of the medical experiments performed at the infamous Dachau prison. This "mission" also included "the wholesale importation of more than six hundred top Nazi scientists under the auspices of Project Paperclip, which the CIA supervised during the early years of the Cold War."

Nazi Rocket Man: Pictured are Pickering, Van Allen, and Wernher von Braun. Hitler once compared von Braun, creator of the Nazi A-4 rocket, to Alexander the Great.

Among the many scientists who came to the U.S. from Nazi Germany was Dr. Hubertus Strughold, the scientist whose chief subordinates were directly involved in the cruel "aviation medicine" experimental tortures at Dachau. Hailed by NASA as the "father of space medicine," Strughold's name later appeared on a list entitled "Reported Nazi War Criminals Residing in the U.S." Despite recurring allegations that Strughold had sanctioned medical atrocities during the war, he settled quite comfortably in Texas and became an important figure in America's space program. After Werner von Braun, Strughold was the top Nazi scientist employed by the American government in the NASA space program.

Another Nazi war criminal who was secretly brought into the U.S. by the U.S. Air Force in 1947 was military officer General Walter Dornberger. Dornberger had been in charge of a rocket factory near Nordhausen, which housed the terrible secret of the history of rocket technology. As Christopher Simpson writes in *Blowback: The First Full Account of America's Recruitment of Nazis*:

> At least 20,000 prisoners–many of them talented engineers who had been singled out for missile production because of their education–were killed through starvation, disease, or execution at Dora and Nordhausen in the course of this project.

Thousands of these slaves in the Nordhausen rocket factory starved to death, and cholera killed hundreds of inmates daily. Cremation of dead bodies in ovens kept the bodies from putrefying and spreading disease. According to Simpson, there were so many dead bodies laying around that the rocket work under Dornberger's direction took precedence over picking up the dead bodies lying about. Simpson writes that General Dornberger visited the rocket factory often, and was well aware of these conditions.

After the war, Dornberger was put to work on a classified rocket project at Wright Field. In 1950, Dornberger went to work for Bell Aircraft where he was a liaison with U.S. military agencies. Dornberger held top security clearance and won the American Rocket Society Astronautics Award in 1959. This is only a brief description of the legacy of the American space program's ties to Nazi war criminals. (See also, *The Nazi Rocketeers* and *Sex and Rockets*.)

Given the details of NASA s Objective 16 discussed earlier, it would certainly appear there is some type of urgency in building the IS(I)S station and ejecting a select few, along with the genetic "seeds" of life, off the ground and into space in a hurry. Yet, I wonder if it is vanity to assume the "gods" are the least bit interested in the propitiations of the kings of Sodom and Gomorrah. In light of the spectacular news that a Jovian-sized planet has been spotted in the outer regions of the solar system, and in light of compelling new evidence that Earth has obviously been hit before by a large Mars-sized body, IS(I)S may very well be a planned Alternative 3-type scenario: a rush to get the elite guard population OFF planet quickly.

Scattered People=Scattered Language=Scattered Mind

The Tower of Babel story represents the loss of human memory, language, and autonomy. A vague concept of "autonomy," to be "like the gods," usually attends this defiant act of the Babylonian people; but is there more to it than that? The first language of mankind was said to be a universal tongue. We can surmise that this one unified language made mankind capable of vast understanding and camaraderie. Indeed, there is some indication that the earliest people may have possessed an incredibly clear communication apparatus, to the point of telepathic capability. But "something" happened on the way to Babylon.

Alien grays and other abductors are said to be able to communicate in any language. What might explain this anomaly? Is there a basic, functional, unified language upon which all human languages are built?

According to Ingo Swann ("Smaller Picture vs. Bigger Picture," www.biomindsuperpowers.com), the ability to communicate is "downloaded from the language factor." The language factor is operative and functional at birth. Swann points to a 1993 *Life* magazine article called "The Amazing Minds of Infants." Presenting an overview of recent research in physics, mathematics, language and memory, this article asserts that from birth to four months, babies are 'universal linguists' capable of distinguishing each of the 150 sounds that make up all human speech. During this period, and before they begin learning words, babies appear to be "sorting through the jumble of the 150 sounds in search of the ones that have meaning." By about six months, they have "begun the metamorphosis into specialists who recognize the speech sounds of their native tongue." Swann comments that this process of sorting through the basic speech sounds of a "local language system" describes language *decoding* more closely than it describes language *learning*. Essentially, what this means is that humans do not learn a language system, rather, they "recognize which language system is going on about them."

The most dangerous power which humans can potentially possess is collective Mind, or telepathy. Psychic Ingo Swann considers telepathy one of the

"Superpowers of the Human Bio-Mind." We might even assume that such a potential can be an inherited factor, just as the language factor is inherited. If a race of advanced beings wanted to alter this collective dimension in their human offspring, they might create a rift or chasm in this collective dimension by splitting the core language of humanity into many tongues. By dispersing the people over the globe, they would eventually lose the power of telepathy, or collective Mind, since they essentially would forget that they ever had it. Seen in this light, the scattering of ancient people and the confusion of tongues preempted the scattering of Mind-at-Large, or humanity's inherent telepathic powers. This was the ultimate act of mind control.

Myth has it that human language incompatibility was effected as part of humanity's punishment for certain "violations," but this act might also be seen in the context of restrictions and boundaries, a divide and rule tactic, and the ultimate act of mind control. For, according to Harry Hoijer in "The Sapir-Whorf Hypothesis," (*Language and Culture*, 1974), "no two languages are ever sufficiently similar to be considered as representing the same social reality. The worlds in which different societies live are distinct worlds, not merely the same world with different labels attached."

The confusion of tongues was the result of humanity's attempt to create an autonomous future, and should be considered within the context of planet Earth being under quarantine. If the assumptions underlying this book are correct, the babysitters of the human race will not allow us to leave the planet. As Brinsely Trench suggested in his 1960 book *The Sky People*, "Planet Earth is under a kind of quarantine. The Sky People will not step in overtly until humans actually project themselves into space so effectively as to become a threat to the galactic order." Although we have been sending out unmanned probes, it is unlikely that we will be allowed to undertake any serious manned exploration of the solar system. There is something out there that does not want us to venture too far. The following story illustrates this profoundly stark realization on a different level.

Birth, Death and Alien Abductions

Many abductee stories include "fetus" or "birth" motifs. Dr. Alvin Lawson's Birth Memory Hypothesis, as outlined in *MILABs*, asserts that many alien abduction cases have a connection to spontaneously remembered traumatic birth memory experiences. This theory asserts that sudden birth memories could be the cause of the alien abduction motif. Dr. Lawson has "induced" imaginary abduction experiences in sixteen volunteers and later recovered the stories. The reports tended to include images such as placental disk-like crafts, fetal humanoids, womb-like chambers, bonding and nurturing motifs, the "tunnel of light," and various tube and tunnel motifs resembling the birth canal or "cervical doorway." Being inside of a UFO has been likened to being born in various ways: including trauma, humming sounds, extreme tastes and odors, breathing problems, head and body pressure, paralysis, body size changes, life "review," amnesia, and time loss.

However, there is good reason to question this hypothesis in light of the fact that a majority of people cannot remember anything prior to age three, never mind "spontaneously" remembering being born. Furthermore, might there be some other explanation for the similarity between UFO motifs and the process of being born? Abductee Lanny Messinger has deduced that alien Gray and Reptilian groups (apparently under supervision of the Serpent Staff Pleiadians) have "intentionally designed their programming methods and equipment to parallel the natural birthing process." Lanny has experienced recall of various past life events, including the recollection of between life "implant stations" and their ET occupants.

In his book *The Programming of a Planet*, and in a Spring 1999 *Paranoia* article, Lanny Messinger outlined the story of his 1995 breakthrough in understanding the ET phenomenon. The breakthrough concerned "the literal and symbolic meaning of the *light* and its use as a hypnotic trigger to control Earthbound humans." Lanny warned that few would be able to accept this information due to the belief systems that have been programmed into Earth societies. He asserted that the "big picture" was one which involved both past lives phenomena and alien abduction phenomena. After researching past lives, Lanny and his wife realized that they had been involved in continual alien abductions for a number of years.

Lanny came to the realization that "white light rooms" were being used by Grays and Reptilians for the purposes of "programming posthypnotic suggestions and installing amnesia" in-between lives. During many abductions into spacecraft, Lanny recalls being taken into "a room filled with a thick, foggy, white light ... so thick that one can barely discern someone standing two feet away (unless they have two big black eyes that stand out in the contrasting white light)." Lanny has also uncovered a past life abduction which occurred in 1942 in which the airplane he was flying was beamed up into a very large Reptilian ship. The ETs pulled him out of his airplane and strapped him into a metal chair, where powerful bursts of white light from "an electronic torture device" drained him of his will power.

In a different room filled with white light, Lanny was inculcated by the Reptilians "to be aggressive, to like war, to want to kill, and that killing gives one power and strength." They told him (telepathically in German) that he was a "Chosen One." They informed him that "they would take care of me, I belong to them, and I'll do their bidding." Lanny was also told that Earth is his home, and would *always* be his home. He was then taken to another white light room and strapped to a table, which was then spun around horizontally. He had also experienced this on alien Gray's ships, and he believes it "is used to disorient the abductee in order to reinforce the amnesia."

As he continued to open past life memories, Lanny began to realize he was being processed in these white light implant stations between lifetimes. The implant stations are "gargantuan motherships that have very large white light programming rooms." After death, he explains, a person "enters the implant station through a tunnel and toward the light at the end of a tunnel." Lanny has noted that some of the alien beings in these stations are "insectoids" resembling a large "praying mantis."

The Light at the End of the Tunnel

Lanny recalls the following scenario: "Up until 12,389 years ago I was a free being. I had been hanging around Earth for thousands of years until I got too close to an implant station. These implant stations have some type of tractor beam—apparently electronic in nature—that can paralyze and pull in a free being. I was pulled into the implant station through a tunnel. I traveled toward the "light" at the end of the tunnel." After coming out of the tunnel and into the "light," he was "electronically compressed (implanted) into a female human body by a device that radiated an overwhelmingly powerful white light force field."

After this experience, he was escorted into a large white light room where other bodies milled about like zombies in a "thick, misty, white light." In this room, he heard soothing, hypnotic commands to "sleep," and "don't worry, we'll take care of you," as well as hypnotic amnesia commands like "you will not remember," "you have no past," "be here now," and most importantly, "always return to the light." In one re-birth episode, Lanny recalls that he left his body and "reported in" at the implant station, entering through a tunnel and traveling toward the "light." Lanny then entered the body of a baby girl on her way out of her mother's womb, via the "tunnel" and toward the light at the end of the tunnel, which was the birth canal.

Lanny has subsequently recovered many similar memories of between-lives incidents. In most cases, he reports in to the implant station as a spiritual being and then bounces back "to a hospital or wherever there is a mother-to-be in her last moments of labor." In one in-between lives incident (1312 A.D. in England), Lanny claims he "rose about a half mile above the dungeon in which I had just been tortured to death, only to find a small, alien Gray's disk waiting for me. They "escorted" me to another area of the country where I entered the womb of a mother giving birth."

Lanny explains that a free being is "a state of spiritual beingness whereby there exists no compulsion to be in a body or tendency to be stuck in a body. A free being is not stuck in the continual life and death reincarnation cycle. A free being can be in a body or not be in a body at will—he's just not stuck in one." In his research, he has "uncovered considerable evidence that implicates an elaborate conspiracy to keep mankind in spiritual darkness." He implicates the Grays, Reptilians, Insectoids, and "Serpent Staff Pleiadians" as the ET groups participating in this conspiracy. He believes that all Pleiadian contacts, including Ashtar Command, Hatonn, Semiase, (whether channeled or not), "are disseminating disinformation and should not be trusted." As a matter of fact, Lanny believes there are no ETs telling the truth.

Lanny concludes that in-between lives inculcation explains why people don't consciously remember their past lives, and that amnesia is a tool used by negative ETs to enslave mankind within the boundaries of certain belief systems. Not everyone is abducted in their physical bodies during their lives on Earth, however, "all the earthbound humans who are stuck in the compulsive reincarnation cycle are sometimes abducted as spiritual beings after death due to this ubiquitous programming." Lanny also concludes that Earth is a kind of "prison planet." It is possible that people can disable their programming by recalling these incidents with past life therapy.

The Earth Custodians

Related to this story, ancient Sumerian writings indicate that the 'gods' wanted to permanently join spiritual beings to human bodies in order to create a slave race. As William Bramley writes in *Gods of Eden*, it would appear that the existence of our spiritual beingness has been "in some way deliberately blocked from human memory by the Custodial Society as part of its effort to weld spiritual beings to human bodies." Bramley further suggests that the Mormon religion contains ties to such a deliberate scheme ordained by the "custodians." Mormon writings, which appear to have been given to Joseph Smith by a supernatural entity, declare that eventually the spirit world would be eliminated entirely, and only the material Universe would exist for the people of Earth. Bramley takes this to mean "total spiritual entrapment in physical matter." In order to effectuate such a condition, he surmises, philosophies of strict materialism would need to be imposed so that humans do not look beyond the material Universe. This condition would necessitate the widespread teaching that "all life, thought, and creation arise solely out of physical processes." Bramley comments that such materialist philosophies have indeed pushed the human race into "an ever-deepening spiritual sleep."

The reason for the widescale peddling of materialist philosophies comes clear in Allen Greenfield's book *Secret Cipher of the UFOnauts*. He writes, "in elevated mystical states, the very secretions, psychic and otherwise, which humans emit and which desperate vampire aliens consume like the soul-famished pathetic creatures they are, become poison to them in the Transformed Human. Cosmic Consciousness is literally poison to them." He explains, far outstripped in terms of physical weaponry, humans must find a source of strength that transcends the physical using the techniques of centered Cosmic Consciousness.

David Icke has written in *The Veil of Tears* (see www.davidicke.com) that certain frequencies can be used to block receipt of radio information, and that the Soviets have experimented with such technologies which can create a "vibratory prison." He asserts that all we need to do is extend this concept to the planet as a whole to get an idea of how spiritual information is being perceptually blocked from the 4th dimension, creating a material prison. He writes:

> The complete takeover of the Earth by extraterrestrial expressions of the Luciferic Consciousness was accomplished by creating a vibratory prison. We are multidimensional beings, naturally able to experience many frequencies and dimensions at the same time. However, when the imprisoning vibration – an imposed blocking 'frequency net'– was thrown around this planet long ago, it prevented us from accessing the higher levels of our consciousness and potential – or, the higher dimensions. It caused us to cease to be 'whole' or 'holy,'and we became disconnected from 'the Father.'

According to Icke, the full-scale vibratory imprisonment of Earth humans may also have been effectuated by closing down crucial Earth vortexes linking the physical (3-D) world with other space/time dimensions. Some of these important windows, however, are still open, and, as well, certain rituals are said to be able to re-open them. Icke speculates that these interdimensional portals may have been closed to prevent negative entities from entering this space/time reality. Thus, this may have been a necessity to minimize chaos and disorder. However, it left Earth humans detached from higher levels of being and cut off our "eternal memory" of who we are. This is essentially the story of the "Fall of Man." He writes, "we forgot who we were and where we came from." Icke explains:

> The human race has for ages been living out its existence inside a kind of metaphysical box with the lid held down. We sit in the dark, believing that our potential, and Creation in general, is limited to what is within that box, within that vibratory prison. Over the ages since the vibratory net was cast around the Earth, we have been a people working at a fraction of our full and infinite potential. Life on Earth was changed dramatically by our extraterrestrial jailors, and this also affected the animal kingdom.

Luciferic consciousness works through human consciousness to manipulate human nature and our understanding of reality. It stimulates us to perform inhumane acts by awakening negative emotions. Icke believes, however, that there are positive 4th dimensional extraterrestrial groups working to help humanity rise up and reconnect with our lost identity. He suggests that the phenomenon of crop circles is an example of positive extraterrestrial contact. But if you think this tale is strange, it becomes even more bizarre up against the following information regarding "the Screen," a programmed vibratory net which has purportedly been encountered telepathically by splinter-group Scientologists.

The Screen

According to L. Ron Hubbard, founder of Scientology, various "implant" scenarios have occurred throughout the eons, causing humans to be stuck in material bodies over and over again. As a result of various implants, thetans became "firmly welded" with MEST and their GE, or Genetic Entity. In this way, matter, energy, space and time (MEST) came to be seen as pre-existing. Splinter-group Scientologists working in the high-level Excalibur project have discovered a "Screen" which has been erected over the planet Earth, which was part of an ancient implant scenario called "Inc.2." The population of the Earth at the time understood that this Screen was being erected as a detection and defense screen, but the Screen had the actual purpose of preventing thetans from leaving Earth and their material bodies. (Note that the Inc.2 scenario seems to be reminiscent of HAARP!)

According to L. Kin in *Pied Pipers of Heaven*, the Screen is a grid of entities (theta quanta) comprised of standing waves containing scalar waves. It vibrates at about 2 Gigahertz and seems to consist of plasma spirals working in opposing directions, providing stability. The Screen is described as having the appearance of a honeycomb, with about 49 grid crossings per square meter. Each cell has six sides and contains a copying and programming mechanism. As Kin explains, the thetan becomes attracted to the Screen because it resonates at a sympathetic frequency. The thetan considers it familiar and doesn't notice it. On contact with the Screen, the thetan forgets his own past and identifies with the information being programmed by the Screen. Since the GE Pool also comprises part of the Screen, the thetan is coupled with a GE, or Genetic Entity, when he approaches the Screen, and is reborn in another Earth body.

Kin writes that high-level Excalibur (unofficial) Scientologists have been acting to counterprogram the Screen since it was discovered in 1992. This is part of the Excalibur project. They have been approaching the Screen telepathically, utilizing a concept much like a computer virus. Focusing on the Screen, the Excalibur deprogrammer drains the energy content of one of the cells. As the cell spills its energy, the Excalibur deprogrammer telepathically charges it with standard Excalibur anti-Xenu information, such as "Xenu's game is over." Since the Screen is actually comprised of theta quanta, the entities are provided the standard "Two Rights of a Thetan," which are the right to self-determination, and the right to leave a game. The thetans are told that they have the right to do whatever they want, and they no longer have to play the Screen game.

Kin explains that during initial trials of this deprogramming process, wave impulses began to spread throughout the Screen, and the copy mechanism with the anti-Xenu instructions began to diffuse in an outward motion from the cell. Several repetitions containing the anti-Xenu instructions seemed to cause the program to run by itself, permeating the Screen. When last checked, the counter-program appeared to be running on its own. It has been noted by auditors that friendly Visitors to Earth now find it easier to move in and out of the vicinity of Earth since the deprogramming of the Screen.

Interestingly, in his travels of out-of-the-body, Robert Monroe has indicated that there is a kind of conglomerate of human "bodies" circling the Earth vicinity. All of this information needs to be taken into account for a full picture of the bizarre nature of reality. This book has attempted to bring into consideration some of the information pertaining to the nature of the Earth as a protectorate for a rare species of humanoid, and this species' biological and spiritual manipulation and entrapment in the Earth's material environment. But there is surely much more information out there to add to this bizarre scenario. As I've said elsewhere in this book, all hands are needed on deck to decipher this bizarre reality.

Invasion of the Body Snatchers

If a condition of amnesia coupled with paradigm thought control were being manipulated from the 'outside,'would it require some type of field command or ground troops? Does this imply that certain factions within government agencies, perhaps with commands coming from "higher up," are either extraterrestrial in origin, or are hybrids, or are working for ETs? This is one scenario that most people are not ready to handle, but it needs to be examined as a possibility. As Philip Corso has expressed, the US military has established a defensive presence with regard to what it has perceived as hostile ET threats, and *maybe* this has kept them at bay, or has at least kept the abductions of humans at a minimum. He asserts that humanity "won its first victory" against a technologically superior enemy who discovered there was "real trouble down on its farm."

It is interesting that Corso uses the word "farm." Charles Fort suggested that humans were "fished." Taking these ideas into consideration, Corso's militarist War of the Worlds posturing ranges from completely immature to ludicrous. It may already be too late. The EBEs already had it figured that if they really want the farm, all they have to do is climb in through the bedroom window. Some "hybrid" human-aliens may be doing the bidding of ET forces, and human abductees may be implanted to do their bidding in the future. It has been reported that a human agency is somehow involved in alien abductions. The suggestion that humans are aiding and abetting in certain incidences of alleged alien abduction is outlined in a new book called *MILABS: Military Mind Control and Alien Abductions*. Military-type abductions, reported primarily in the U.S. and Canada, mimic the alien abduction scenarios without the reported supernatural motifs. Instead of UFOs, they use unmarked helicopters, vans, and buses in order to kidnap victims and bring them to underground facilities. Whereas alien abductors appear through a closed window or a wall, and paralysis occurs via telepathy, MILAB abductions induce compliance with a syringe.

The MILAB theory suggests three hypotheses: (1) covert military and intelligence agencies are carrying out mind control projects under cover of "alien abductions," (2) covert agencies are carrying out clandestine biological and genetic engineering research in underground facilities under the "alien" cover, (3) a covert task force is monitoring real alien abductees in order to debrief the victim, as well as to install full amnesia regarding both abduction incidents. In either case, covert units seem to be up to "something extremely strange and secret hidden behind a criminal veil of kidnapping and narco-hypnotically induced mental screens." The agenda, points out Richard Sauder in the book's foreword, may be "dark and occult."

It must be conceded that some "alien" abductions might be mind-controlled Virtual Reality Scenarios pulled off by Earthling reality engineers. This explanation has quite a following among UFOlogists, including Norio Hayakawa, William Lyne, and others, who believe that a secret cabal is preparing to pull off a holographic hoax in the skies. Skeptics point to these covert 'military maneuvers' as evidence that most alien abductions are "staged" by the intelligence community as a cover-up for their illegal mind control and genetics experimentation. This theory suggests that military intelligence is field testing brain implants, virtual reality implants, holographic image projection, cloaking devices, mind altering weapons, and genetic research.

The above may be a somewhat simplistic explanation, however, since the MILAB-type abductions represent a *fraction* of the information in abduction research files. For those of us who are unable to get around the facts of the historical presence of ETs, as well as the non-Western incidences, there is another possibility being left out of the picture. Could it be that alien-human hybrids have already infiltrated ground forces on Earth? Dr. Jacobs, author of *The Threat*, suggests that MILAB abductees "mistake human-like alien hybrids dressed in military-like clothes and uniforms for U.S. military personnel." He suggests that the military association may be a ruse pulled off by alien-human hybrids.

MILABS suggests an even more bizarre counter-plot. This potential scenario suggests that MILAB abductees are, in effect, "trojan horses" inside "real" UFOs. The interest shown by military intelligence indicates that alien abductions are real. As *MILABS* suggests, MILAB abductees appear to be alien abductees who are being alternately programmed to work for the military as "hypnotic couriers." A covert task force appears to be involved in tracking these couriers, as well as hypnotically controlling them to obtain information about the ETs and their abduction experiences. The mind control victims are entrained to "dissociate" their personalities, so that the "spy" personality is buried beneath the victim's normal personality. The victim is not aware that he or she is working as a spy for the covert government. This type of "Manchurian Candidate" has been historically traced in the mind control literature, but it has become evident that there is a new use for these mind control victims in a ground war with telepathic extraterrestrials. The psychic factor may also come into play, since alien abductees appear to be telepathic, and the secret government is certainly interested in the human telepathic capacity.

The issues of mind control and alien telepathy interweave in the larger picture that is coming into focus. New information suggests that psychotronic weapons are being developed for a purpose, and that purpose is a ground war with telepathic aliens. It may also be that exotic technologies like HAARP, SDI, and a host of electromagnetic, atomic and particle-beam weapons are being deployed in a last ditch defensive effort. On the other hand, if we consider the concept of the "Screen" and how it was erected ostensibly as a defensive shield, what might be the alternative uses of HAARP as an electromagnetic fence for keeping us animals inside of our cage?

There is also the sense that this covert task force itself may not be entirely human, and planetary security may long ago have been breached. In either case, a terrestrial War of the Worlds appears to be underway. The ultimate question is: how long has this been going on? Have Men in Black now become Men in Black Helicopters? And, are they of "Earthside" or "Spaceside" origin? This question has not changed over time. Neither has it been answered.

In any case, I would not suspect that the EBEs feel seriously threatened by us, as Corso would have us believe. I suspect that the EBEs do not want us to venture too far in the exploration of our solar system, due to our dangerous emotions and aggressive tendencies. It will be interesting to see whether the ancient Custodians allow us to build our International Space Station, even though NASA has taken the trouble to make ritual offerings to the ancient Watchers of humanity.

The Project

In his book *Angels and Aliens*, Keith Thompson has suggested that the real "anomaly" is "modern Western culture's *project*—unparalleled in history—of placing entire dimensions of reality off limits, thereby forcing these dimensions to 'return with a vengeance' in various forms, including UFOs." Perhaps Thompson did not mean to emphasize this, but I think his choosing of the word *project* should not be underestimated. This modern anomaly is an extremely pervasive and peculiar *project*, and is difficult to get a grip on, since the other "P" word also applies. You will certainly be considered *Paranoid* if you broach this subject among people who rather enjoy sitting in the dark. As Key writes in *The Age of Manipulation*, there will always be "dissidents, disbelievers, deviants, heretics, subversives, and critics who struggle against mainstream conditioning." But be sure that these deviants are watched carefully. If they attempt to overhaul the system by alerting people to the mechanisms of control, Key warns, they will be "discredited by the majority and its power elite."

Key also points out that conscious memory and perception is limited by avoidance behaviors. We see what we want to see, through the filter of our symbolic reality. If it doesn't match up to this reality template, we deny its existence. As Key asserts, we habitually deny what has been excluded from our limited perceptual conscious awareness. But who or what has limited our awareness to a short range of perception? The answer begins with our saturated ad-media environment. But this analysis only scratches the surface.

We also need to look at what our science and technology tells us about our environment, about our world and our place in it. Key reminds us that "the moment a scientific fact is cited in support of an argument, it no longer has anything to do with either science or facts ... science becomes psychological silly putty when used to hype an industry, cause, ideology, product, person, group, or nation." In high-tech cultures, science becomes a "mythological, godlike creation presumed omniscient, omnipresent, and omnipotent." When "scientific truth" is accepted without question, it terminates the intellectual process. Key warns that "unquestionable scientific facts attributable to high-credibility sources impose an end to critical thought." As Key warns, "science must always be viewed as tentative."

Could powerful Earthling forces be simply mimicking on a small-scale what their Overlords have done on a large scale? The manipulation of mental constructs may be a mini-design of a much larger scale phenomenon: the engineering and control of large physical constructs in the solar system. The satellites of Mars and Saturn may be huge terraformed engineering projects put in place by advanced spacefaring ETI. Certain information that should be widely known is kept at bay by means of the "giggle factor." The reason we can't even have this discussion openly is due to the sociological phenomenon of paradigm thought control which underlies this giggle factor. This type of mind control is put in place by a rigid scientific paradigm which is only capable of seeing the human existential situation from the inside-out rather than from the outside-in. This paradigm control will become more obvious after reading the chapter on evolution and beginning to pay attention to terminology which is aimed at keeping you convinced you came from the "water" rather than the "sky."

Humans do not like uncertainty, so they invent rigid "factual" structures or accept media-induced reality structures which are basically designed in terms of black and white opposites. But it is important to realize that such either/or verbal dilemmas are only word manipulations. You are not required to engage in this distorted dialectic which boils down options to logical oppositions: yes or no, good or bad, true or false. You have an endless play of options, because you are a human being, a gray area, an event pending, a connecting link. As Dr. Key writes in *Age of Manipulation*: "uncertainty might be a great source of endless human joy, innovation, creativity, excitement, multiple perspectives, and the challenge of intellectual adventure."

Perhaps there is, in Goswami's words, a potion that can sever the quantum self. It is utilized by the military-industrial-media-entertainment-academic-scientific-complex, and it constitutes a worldwide information management conspiracy. If we try to tune out this pervasive influence, we could at least curtail our most immediate and obvious mind control, and free up our perception in order to see the total psycho-spiritual flummoxing which exists beyond it.

The Anthropomorphic Bias

In response to E.J. Coffey's statement discussed earlier, that the scientific view of the improbability of humanoid intelligence elsewhere is "not due to any anthropomorphic bias" but is due to "the scientific bias [that] sees evolution as adaptation to specific local environments," I would argue that this conclusion *most certainly is* based on an anthropomorphic bias. The "Face" on Mars is a prime example of our own anthropomorphic bias. The simple fact that we cannot allow ourselves to believe that humanoids resembling ourselves ever walked on a planetary body in our own solar system is proof of our anthropomorphic bias. And this bias is based on our current "scientific bias," which sees human origins as a purely local phenomenon.

A confounded dilemma trips up the use of evolutionary theory as an argument against the co-existence of the humanoid form in the Cosmos at large. The argument is essentially confounded by the fact that we do not actually *know* that we are the only humanoids in the Universe, nor do we *know* the genesis of the humanoid form. Therefore, we are extrapolating this information from an anthropomorphic bias: the theoretical "random" evolution of the human form from the great ape lineage as a purely localized anomaly. The argument is circuitous. After all, wouldn't the appearance of an ET race in our skies automatically make short work of Darwinian evolution?

The Space Travel Argument *Against* the Existence of ETI argues that the Zoo Hypothesis is unlikely. If it were true, they state, "our entire Solar System would be analogous to an American national forest, or an African game preserve" and von Neumann probes would be presently acting in the capacity of "Game Wardens." If our Solar System were such a preserve, they surmise, "then all contact must have been rigorously prevented for as long as the robot game wardens were present in the Solar System, since there is not *one jot of evidence for any contact in the past*." Such a total prevention of contact is, they assert, quite impossible. Surely someone would *try to get through*; some ETI group would believe contact to be the interest of Earthlings.

Our "lack of detection" of ETI life forms could just as well be an indication that an "embargo" is in place. Such an embargo, or quarantine, of Earth humans could be effected by means of telepathic or technological mind control, as well as strict philosophical and religious paradigm control, instituted by either Spaceside or Earthside powers, or a combined onslaught. As the Space Travel Argument presumes, wouldn't someone try to get through somehow, any way they can, in an attempt to show us that there is *something more* than material substance? Is this the simple, straightforward message being *beamed* upon us? Are some of the more benevolent *terra intermediaries* trying to get a message through the electric fence of our preserve? Are they trying to rattle the cage of our materialist paradigm?

Chapter Four

The Cyborg Hypothesis:

Smart Probes, Plastocene Portals and the Anatomy of an Alien Gray

A living being that does not engage in a continual exchange of energy with its environment is simply unthinkable.

H.V. Ditfurth
The Origins of Life

One of the most persistent arguments against the reality of what has been termed the Visitor Experience is based on the presumption that if the incrementally accidental and unguided evolution of the human form occurred on Earth, it is nearly mathematically impossible that Visitors arriving from another system, or even within the solar system, could be "humanoids." This idea is made up of two common assumptions: (1) based on an overtly similar morphological design, the most commonly described abducting entity called the "alien gray" is inarguably a "humanoid," and (2) that the accidental and unguided evolution of the species *Homo sapiens sapiens* from the great ape line is inarguably a fact. Let's address the first assumption now and the second later in the book.

The Roswell Crash

It all started with the reported discovery of the remains of a downed saucer on July 3, 1947 by rancher Mac Brazel on his ranch near Roswell, New Mexico. According to the press release dated July 8, 1947, the "disc" was picked up by Major Jesse A. Marcel and pieces of it were inspected at the Roswell Army Air Field and flown to "higher headquarters." This news release, distributed by public information officer First Lt. Walter Haut on the direct orders of base commander Col. William Blanchard, was available to every newspaper and radio station in the country. The find was specifically referred to as a "flying saucer" and a "disc" and news reports implied that it was not of Earth origin. *The Roswell Daily Record* ran the story under the title "RAAF Captures Flying Saucer on Ranch in Roswell Region. No Details of Flying Disk are Revealed."

On the following day, July 9, 1947, *The Washington Post* ended the tale of the crashed flying saucer with a complete reversal after the military changed its story. The "flying saucer" was now a weather balloon. As Stanton Friedman writes in *Crash at Corona*, "no one thought to question why something so commonplace as a weather balloon had caused so much commotion, or how two officers of an elite AAF unit could possibly have failed to recognize it, or how this flimsy contraption [described in the second press release as "tin foil and broken wooden beams"] could have strewn its pieces over a square mile of sheep ranch."

A few days after the discovery by Mac Brazel, military air craft combing the area purportedly discovered more debris a few miles away, along with three or possibly four bodies in some kind of 'escape capsules'. Stanton Friedman proposes that the wreckage and the bodies were gathered up from the various sites and carried by B-29, C-54 Skytrooper and B-25 Mitchell bombers to Wright Field, Ohio and to Fort Worth, Texas, for examination. According to *Truth About the UFO Crash at Roswell*, Maj. Jesse Marcel placed some of the debris on General Roger Ramey's desk upon its arrival at Fort Worth. The two men then stepped out of the office to look at a map of the crash site. When they returned, the original debris was gone, and in its place were pieces of a destroyed weather balloon. Marcel was ridiculed at a later press conference for "being such an ass for bringing that weather balloon all the way from Roswell."

Crash Test Dummies

Almost exactly fifty years after the Roswell hoopla, the Air Force announced their official explanation of the incident. On June 24, 1997, Col. John Haynes USAF read a prepared statement before a press conference at the Pentagon. According to *Uncensored UFO Reports* (7/98), he announced that the Air Force was "confident once the report is out and digested by the public that this will be the final word on the Roswell incident." He explained that the debris recovered near Roswell fifty years ago was the remains of a formerly top secret Army Air Force research project called Mogul. Using an odd assortment of equipment, he explained, Mogul was carried aloft by a balloon train over 600 feet long in an attempt to acoustically detect Soviet nuclear blasts and ballistic missile launches. The 25 balloons making up the balloon train were attached to acoustic devices and radar reflectors that look like box kites. The Air Force asserted that the misidentification of the radar reflectors were most likely the cause of the flying saucer reports.

Col. Haynes explained that these Air Force activities occurred over a period of many years, and vaguely attempted to establish an occurrence of testing "in two or three days in July of 1947." He explained that the bodies observed in the New Mexico desert were "probably test dummies that were carried aloft by U.S. Air Force high altitude balloons for scientific research." He reported that the unusual military activities which are reported to have occurred at Roswell could be explained as recovery operations for this top secret high-altitude balloon research. Haynes explained that reports of military units that "always seem to arrive shortly after a crash are accurate descriptions of Air Force personnel engaged in dummy recovery operations."

Col. Haynes explained that the claims of bodies at the Roswell Army Air Field hospital were most likely a combination of two separate incidents: a KC-97 aircraft accident in which eleven Air Force members lost their lives in 1956, and a manned balloon accident in which two Air Force pilots were injured in 1959. In addition, he stated that the dummies were first used in balloon research 1953. When asked how these dates can be reconciled with the 1947 Roswell incident, Haynes replied: "I'm afraid that's a problem we have with time compression." Haynes asserted that "there were about 2,500 balloons launched during a thirty year period in New Mexico alone, and there were lots of dummies dropped." He stated that, "when people talk about things over a period of time, they begin to lose exactly when the date was."

Given that over a period of time dummies were dropped all over the desert area, Haynes asserted that, " I think it's logical to assume that the people saw Air Force ambulances come out. They saw gurneys come out. They saw body bags come out because the dummies were put in body bags to protect them. They saw people in pith helmets. They saw people in shorts out there brushing the bushes looking for remnants of the balloons. And when you put all that stuff together and spin it, you find that it fits perfectly with many of the occurrences in Roswell during that era." He argues that "if you overlay much of their claims and you look at the Air Force scientific research, you can see it's obvious that's what we're talking about at Roswell."

Strangely, upon questioning, Col. Haynes stated that the Project Mogul balloon tests were never classified and that the "dummy tests got a lot of media attention" at the time. If they were never classified, why did it take Air Force dummies so long to "put all this stuff together and *spin* it?" Why would people report seeing small "alien" bodies and massive clean up operations in 1947 when the Air Force claims it didn't start using dummies until 1953? How could people be off by ten years in their otherwise very detailed recollections of such an important event? How could people "lose exactly when the date was" when the dates are corroborated by numerous witnesses and the stories are published in newspapers?

Does the Air Force take us for dummies? With this vague and ridiculous "time compression" explanation, the Air Force insults our intelligence. It is clear that by "overlaying" witness reports on top of their own "scientific research," the Air Force is attempting to kill two flying objects with one media spin. While supplying "the final word" on the Roswell incident, the Air Force is at the same time establishing a cover story for other credible reports of crashed disks spanning a crucial thirty year time period well known for reported saucer crashes and credible UFO sightings. Are they working overtime to discourage the belief in extraterrestrial visitors since they are powerless to curtail such visits, which includes the night stalking of human and animal DNA?

The Physical Proof

The most frequent argument against UFOs is the lack of physical data, but it is indeed peculiar that when you give some people fairly credible proof, they still choose to accept the government's stance on UFOs. With their track record for lying, why should we believe government stooges over the vast number of eyewitness reports of UFO sightings indicating that UFOs might be of extraterrestrial origin? The Roswell crash left behind some very important physical data; including possible extrabiological entities (EBEs), panels containing strange hieroglyphic symbols and hand prints, and materials unknown to Earth. If the Roswell material was man-made, why wasn't it offered to the public for inspection at the time?

If the materials confiscated at Roswell were of extraterrestrial origin, and if our government was run by sane individuals, these cosmic gifts would be in a museum for all to see. Instead we are left holding an empty bag. We have nothing but the word, and the word alone, of witnesses who have been warned to remain silent on the matter. The lid was slammed shut on the Roswell UFO crash since there had been no 'smoking gun' or physical data to prove it had happened the way witnesses described it. But family members and neighbors of the rancher who had handled pieces of the wreckage knew that these materials were unlike any Earth material they had ever touched. For instance, the metallic "tin foil" material was extremely lightweight, but could not be damaged in any way. There was also some threadlike "silk" material that was as strong as wire. In a talk published in *Fortean Times*, (1/96), the son of Major Jesse Marcel, now a physician, stated that the "I-beam" pieces he handled as a youngster had "strange hieroglyphic symbols on the inner surfaces in pink and purple."

The granddaughter of Sheriff Wilcox, interviewed in 1991, stated that her grandparents were warned that their entire family would be killed if they let out any details about the incident. She states that four 'space beings' with large heads were found at the site, and that at least one of them was alive. This statement is corroborated in a saddening disclosure by Linda Moulton Howe in an Internet UFO Newsletter *ISCNI Flash* dated October 1, 1995. She states that she met Ray Santilli, the owner of the Roswell autopsy film, at a convention in Italy. Santilli stated that the cameraman told him that the three broken "control" panels shown in the film, which also exhibit mysterious symbols and hand prints, were clutched tightly by the EBEs to their chests when they were found. The cameraman told him that "a firm strike at the head of a freak with the butt of a rifle" was how they pried loose one of the incredible panels.

The Anatomy of an Alien Gray

The Cyborg Hypothesis has entered provocatively in the UFO literature over the past few years. Let's outline the information acquired to date. According to a talk given by Dr. David Jacobs of Temple University, entitled "The Anatomy of an Alien" (Omega Conference, 10/88), a wide variety of alien types have been witnessed first-hand by sane people all over the world, but, Jacobs adds, the wide variety of descriptions may have more to do with a difference in perception, interpretation or reporting. According to Jacobs, the overt morphology of the most enigmatic of the Visitors, the Grays, falls within a range of humanoid features, but under close scrutiny these reports describe a radically different creature "under the skin."

On the surface, Earth humans and alien grays have a comparable number of appendages. It is true, agrees Dr. Jacobs, that the grays don't have several eyes; they have two. Neither do they have several heads; they have one. They are not "polka-dotted," he observes. They have two arms and two legs, and they walk upright. Biologically speaking, explains Jacobs, for locomotion only four legs beats two legs; but bipedalism frees up the hands. We all know what opposable thumbs are for; but it is not clear that the grays typically have them. They have been variously described as having three, four or five fingers. Two eyes are better than one, and locating them at the top of the head so they are directly connected to the brain makes visual reports instantaneous. As a model for survival, Jacobs suggests, the bipedal humanoid appears to be an adaptive and efficient one. Although Darwinists claim to know otherwise, let us establish the simple fact that no one *really* knows how old this model might be, nor its true genesis.

Conversely, there are important differences which are not immediately noted with respect to the alien gray, since our expectations of a being from another world based on our science fiction stories might be that of a drastically different order. While the skin color varies with reports from whitish or off-white, beige, tan, various shades of gray, and sometimes a green or blue tinge (probably caused by artificial lighting), one of the most important observations is that *variations in skin tone do not exist*. Jacobs points out that the skin is remarkably smooth and even-toned, exhibiting no discolorations, bumps, blemishes, bruises, freckles, creases, wrinkles, pores, hair or hair follicles, veins, etc. The skin is often described as being like *plastic* or *rubber*. Importantly, there are no veins visible which would indicate blood flow beneath the skin. There is an "odd simplicity as if it is a *manufactured* or *synthetic* skin stretched over the frame," he reports.

The shape of the alien gray's head has been described as a "parking meter" or a "light bulb." It is somewhat human-like but "very alien." The large dome or cranium is entirely hairless; there is no hair, fur or fuzz anywhere on the head or, for that matter, on the entire body. There have occasionally been striations noted on the head. Notably, Jacobs points out, communication is not dependent on facial cues. Abductees report that their alien abductors were "happy" or "excited" about something they did; but these expressions are portrayed telepathically. The facial "muscles" do not register emotion. It has been reported, however, that the grays may change the shape of their eyes, or move them slightly, to register an emotional cue. Therefore, although the face has a *some - what human* morphology, he asserts, the underlying musculature which might portray *meaning* or *emotion* is missing.

With respect to the large, black "wrap-around" eyes of the alien grays, there are several interesting features. There is noted to be little or no movement of the eyes, and there is no separate pupil or iris area in the eye. The eyes have functions beyond sight in that they are apparently used for telepathic communication. The eyes do not move from side to side, and there are usually no eyelids seen. Some abductees have seen their abductors blink their eyes in a manner resembling the shutter of a camera lens, but it is not a common report, Jacobs explains. It would seem that there is no need to cleanse or protect the eyes. There are also no eyebrows, although a completely hairless ridge is noted above the eye. There are no ears. If it doesn't need to talk, apparently it doesn't need to hear.

Some abductees report seeing a small mouth, but most report just a slit with no lips just above a very pointed chin. Jacobs points out that there is no evidence of a "hinged jaw" or muscles to open the mouth, nor are there teeth noted in the mouth, indicating that the beings do not consume in the manner typical of Earth humans. Mastication appears to be unnecessary. Interestingly, early abductees Betty and Barney Hill reported that there was actually a membrane covering the mouth.

Some alien abductees describe a very small nose, and others note just two small holes in the area where a nose would be expected. Abductees are sometimes not sure if they actually saw nostrils or just expected them to be there after the fact. A nose would be useful for many reasons, at least in the Earth environment. Jacobs points out that the nose is used not only for taste and smell, but is also a 'heating chamber' for the lungs in cold climates. Nose hairs also trap pollutants. What might these various features taken in combination say about the alien environment? According to Jacobs, we might deduce that the alien environment is "highly controlled, uncontaminated, unnatural and highly technical."

The alien gray's neck is like a thin tube. A twelve-year old boy reported to Dr. Jacobs that he grabbed one of his abductors by the throat. He stated that the neck felt solid, not soft. Drawings by abductees indicate that the neck does not gradually fit into the base of the skull, nor does it gradually slope out to fit the shoulders, but is rather like a straight tube between the head and the shoulders. The chest is small and narrow, with no noticeable clavicle, sternum, or bifurcated structure in the upper torso. In other words, the chest is not divided into two halves. There are no breasts, breast muscles or nipples noted. The chest is *motionless*; there is no motion indicative of breathing.

The upper torso of the alien gray is said to be proportionate with the waist or hip area; the torso is proportionately the same all the way down. There are no apparent hip or pelvic bones. The abdomen area is totally flat, indicating that there is no food being processed in the "stomach" area. There is also no indication of ribs; and there appears to be no male genitalia. Differences between the sexes of the alien grays are not overt; abductees simply report that they "know" or have a "feeling" that their abductees are a male or a female. On the alien gray's back side there is no indication of shoulder blades, spinal vertebra, or buttocks.

The arms of the alien gray are frail and spindley, and there is a "bend" where the elbow should be. The fingers are long and thin, generally with rounded pads or bulbs at the end, while some are reported as being tapered at the end. There are usually four fingers and a thumb, but sometimes three fingers and a thumb are noted. The legs are short and bend at the knee and the feet appear humanlike although they are usually covered.

In addition, the alien beings apparently do not age over time, and are *very curious about the aging process of Earth humans*. After a lifetime of abductions beginning at five years old, Jacobs reports that a 35-year old abductee has changed tremendously. She has been abducted by the same entity throughout her lifetime. While she, of course, has aged, the entity looks exactly the same. He has not aged; there are no wrinkles or age spots on his skin. There is no sense of change over time.

Although they appear to be weak and frail, the grays are reported to be surprisingly strong, and are apparently dexterous enough to dress and undress their abductees. Although musculature is not noticeable, the grays appear to be able to "carry" people; albeit some sort of mind control technology may be aiding them. These entities do not appear to walk, but rather, they seem to "glide" or "float."

Thus, the insistence on the evolution of the human animal as a novel cosmic occurrence has actually not been breached by the appearance of this overtly similar entity since, with the exception of a few obvious morphological features, these beings are actually "very alien." All signs indicate that they do not age, they do not inhale or exhale, they do not ingest food or water, they do not excrete the waste products of food or water, they do not procreate, and they do not hear or talk. As H.V. Ditfurth writes in *The Origins of Life*, "a living being that does not engage in a continual exchange of energy with its environment is simply unthinkable."

Jacobs stresses that we cannot make the presumption that these are biologically living entities in the sense that we understand the concept of 'aliveness.' Neither can we say for sure that they are manufactured beings, although they very well may be. Perhaps, he concludes, they are some combination of the two: biological robots.

The Roswell Alien Autopsy

Interestingly, this conclusion was arrived at separately some years after Jacobs made these comments at a 1988 conference. The worldwide showing of an alien autopsy on Fox TV on August 28, 1995 placed 'aliens' on center stage after decades of media silence on the subject. If you were lucky enough to borrow a videotape without the 'talking heads' who poked their opinions into the Fox Television version, you got to see a lot more of the original footage. The autopsy film was purportedly purchased by British businessman, Ray Santilli, from an 'anonymous' cameraman who claims to have taken the footage for the American government in 1947.

The mysterious circumstances surrounding the film's arrival on the UFO scene merits serious scrutiny in itself. The film footage appears to have been dissected as much as the 'alien' being, since parts of the footage were sold to Volker Spielberg, Ray Santilli's German financial backer. Spielberg has rescinded early offers to show his sections of the footage, which are rumored to be much more disturbing than what we've seen. Somewhere, though, there must exist a complete copy of the most scrutinized film since Zapruder captured the killing of a president in 1963.

Anachronistic concerns regarding objects seen in the film were addressed by researchers early in the investigation. Bell Telephone ascertained that they did make a "curly" telephone cord at the time this film was supposedly shot. The round clock seen in the film was most certainly a common object in the forties. The camera, film type and reel length are consistent with what might have been used by the military in 1947. The Filmo camera that was used would have had similar focus problems, since up-close focusing is difficult without through-the-lens viewing. Kodak representative Lawrence Cate has ascertained from a small piece of a print, leader only, that the codes on it were applied only in the years 1927, 1947 or 1967.

Investigator, Graham Birdsall argues that Kodak never physically examined the film, but merely indicated that the edge numbers were applied in those years. However, Robert Shell, a consultant to Fox Television for the Alien Autopsy television special, states that the film shows no signs of "fogging." Fogging would have occurred on this type of polyester film if it had not been processed within a few years of manufacture, suggesting that it is unlikely the event was staged recently and shot on 1947 stock. (*UFO Magazine*, 11/95) A second argument against the film being staged in 1947 is the advanced special effects which would have been necessary to create this incredibly lifelike EBE.

Later nicknamed SUE, for Santilli's Unidentified Entity, the Roswell female EBE caused quite a brouhaha with cries of it being a doctored-up human corpse or a human radiogenic mutant with severe hydrocephaly (water on the brain). A search in the files of the Institute for Radiation Research at Hiroshima and Nagasaki indicate that nuclear radiation causes random malformations, burns, diseases of the nerves and birth defects in the offspring of those affected. Those birth defects would not consist of the addition of symmetrical appendages; that is, one extra well-formed digit on the hands and feet on this six-fingered EBE. Nor would it cause water on the brain.

There were a few other pictures of the "Roswell alien" floating around which exhibited different physical characteristics. Speaking at the International UFO Congress in Mesquite, Nevada in late 1995, author/researcher Michael Hesemann presented the theory that the six-fingered entity being dissected on the film is one of the entities found in the May 31, 1947 saucer crash near Socorro, New Mexico, northwest of Roswell. Hesemann believes that the discrepancies in the eyewitness reports can be attributed to the fact that there were two crashes in the same vicinity at about the same time, but involving "different species" of EBE.

Photos of another entity show a smaller being with four fingers instead of six, which is more typical of the alien "gray" described by abductees. Another picture which made its way to China and turned up on the Internet was discovered to be one of the models from the made-for-TV movie *Roswell*, starring Kyle MacLachlan, which had been loaned to the UFO Museum in Roswell. It too had four fingers instead of six. (*UFO Magazine*, 1/96) Interestingly, Col. Corso writes in *The Day After Roswell*, that the aliens had six fingers.

The alien autopsy film shows a small, hairless, six-fingered entity with an enlarged cranium and somewhat bloated stomach, perhaps due to body decomposition. The entity is lying on its back on a table while doctors, completely covered in white space suit-looking garb, perform a "routine" autopsy. With precision cutting instruments, they cut the skin and peel it back in a realistic manner, laying bare a body cavity filled with unfamiliar organs which are removed one by one. Doctors who have viewed this film claim that these organs are not comparable to the size, shape or locale of human body organs, and have no human counterpart. Nationally-known pathologist Cyril Wecht stated that the body is like none he's ever examined and called for a full-fledged scientific observation of the film with respect to the morphology of the body and the organs.

Plastocene Portals and Looking Glass Eyes

There are, to say the least, several disturbing qualities exhibited by the specimen named SUE. First, as the autopsy continues, the 'peel-off' black eye 'coverings' are removed to show white eyeballs set deep into the sockets with no apparent pupils. At one point we watched in amazement as the EBE blinked. We went over and over the same section of the video, and even slowed it down, and we were sure it blinked! Another disturbing aspect of the Roswell alien is the absence of a belly button, the insertion point for the umbilical cord which connects a mammalian fetus to its mother. Also disconcerting is the lack of nipples on an otherwise muscular chest which indicates presence of mammary glands. (Note: the torso of the Roswell alien seems larger and more muscular than the typical description of alien grays gathered by Dr. Jacobs.) These details would certainly rule out objections that the cadaver is a doctored human corpse, but they also indicate that the entity did not "come into the world" in the manner in which mammals are born.

How can a "humanoid" with physical characteristics as close to our own not share the basic characteristics of mammalian reproduction? There are a limited number of options: (1) it was "hatched" from an egg or larvae and is insect, reptilian or amphibian in origin, or (2) it is bio-engineered and, therefore, it is a cyborg; or (3) the alien grays are a combination of the above.

The third most bothersome aspect of SUE's autopsy is one which I have not seen myself, but is hearsay. I have seen the video in its entirety a few times and it is not clear where a crystal-like object is removed from the brain. This would be another indication, taken along with the previous two, that the entity is bio-engineered.

Perhaps the sections of the footage owned by Volker Spielberg contain this enigmatic detail. It was reported in *UFO Magazine* that several "still frames" of unreleased footage were shown at the International UFO Congress in Nevada. These frames are very detailed and are supposed to be released in the future. Perhaps these are the frames which purportedly show the brain crystal. Stanton Friedman points out in *Top Secret/Majic* that he was expecting to see something that he did not see in the film. He had heard talk of an autopsy of a second alien, one "without a gash on the right thigh, which others had said was on the tape." He reports that, "one story is that Santilli sold this to a collector to make some quick money."

All of the above qualities—large 'plastocene portals'(which may blink after death and which appear to be removable), lack of a belly button and nipples, no stomach or intestines, and possibly a brain crystal—combine to suggest that the Roswell alien is a test tube baby, a biological robot engineered with real protein-based tissue, flesh and 'blood'and a 'power pack' possibly set to hundreds of years, barring any unforeseen space/time accidents. This level of cloning technology is obviously very futuristic according to our level of understanding, but there are those who believe that clones and hybrids are a very important aspect of the UFO conspiracy. The strangest abduction stories are the ones that tell of underground laboratories where fetuses are kept in liquid tubes. The fetuses are described by abductees as being human and alien hybrids. Someone, somewhere has this technology down to a 'science'. But who is in charge of this cloning project?

Friedman has concluded in *Top Secret/Majic* that the public was hoodwinked and the Roswell "alien" was, in fact, a hoax. Friedman has had a very difficult time getting Ray Santilli and others to corroborate the factual evidence of this case. He has made several important comments with regard to the authenticity of the film. Since the film is not continuous, he asserts, we don't know what took place when the camera wasn't running. In addition, nothing in the film linked the entity to the Roswell crash. Oddly, the surgeons in the film did not weigh or measure the organs as they removed them. Furthermore, no alien abductee has ever seen an alien gray with six fingers; the usual number is four or five. Friedman asserts that the Turner's Syndrome explanation is possible, except that this disease only strikes females, and this entity had no breasts. With regard to the segment of the tape which displays some type of beams with hand prints for six-fingered hands, Friedman comments: "these would not have been cheap to manufacture."

Some might argue that the non-stop barrage of television specials on the subject of extraterrestrials and abductions is simply fulfilling consumer demand. After all, sensationalist media eats up corporate sponsors like so many TV dinners. One of the important questions to keep in mind when looking into the matter of alien bodies is that, hoax or not, the time is ripe for things of this nature to come out into the open. SUE will be known as the Pandora's Box of the 1990s. If the powers-that-be did not want the film shown to a worldwide audience, it would not have been shown, period. The mysterious Roswell cameraman had 48 years to sell his attic memoirs. Why is now the time? What is the message we are to construe from this?

The same message being drilled over and over in UFO circles boils down to two choices: 1) Aliens are real and the world government is slowly preparing us for that truth; or 2) the alien shenanigans is a secret government ruse utilizing advanced psychotronic mind control and cloning techniques, the purpose of which can only be imagined. If the Roswell Alien Autopsy Film was a hoax, did the hoaxers deliberately try to lend credence to the Cyborg Hypothesis?

The Cyborg Hypothesis

The Cyborg Hypothesis was again supported by the revelations of a former Pentagon official in 1997. According to retired Col. Philip J. Corso in his book *The Day After Roswell,* the technological level of Earthlings was immediately catapulted into the future after the crash of an ET disk, which may have been hit by lightning, in the desert of New Mexico in July of 1947. In his book, Corso talks about his assignment to the U.S. Army's Foreign Technology desk from 1961 to 1963. He describes in detail the contents of a file cabinet which was placed in his possession by his superior, General Trudeau, during this time.

Corso's file contained a two-piece set of elliptical eyepieces that were "as thin as skin." A file description accompanying the lenses, written by pathologists at Walter Reed, indicated that these flimsy attachments were adhered to the lenses of the ET's eyes and "seemed to reflect existing light." The pathology report stated that when the researchers peered through these plastocene portals in the dark, the outer shapes of objects were illuminated in a greenish orange hue. Also among these alien artifacts was a dull, silvery swatch of metallic fiber which bounded back to its original shape when folded or bent, and could not be cut with scissors. There were also plastic-like wafers shaped like oyster crackers about two inches in size. The thin, matte-gray wafers, in varied round and elliptical shapes, had tiny "road maps of wires barely raised or etched along the surface." Corso determined that this was a stacked wire circuitry of some kind.

Corso's job was to examine the materials contained in this secret file, which consisted of several bizarre technological artifacts of obvious "foreign" origin, along with the autopsy reports of several dead extraterrestrial biological entities (EBEs) examined by Army pathologists at Walter Reed Army Hospital in 1947. Corso's mission was to suggest the course which the Army should take with regard to these alien technologies. Corso ultimately suggested that the Army diffuse this technology into the military and public domain over the course of time, under the guise of foreign technology, while keeping the general public unaware of its extraterrestrial source. In his book, he outlines his proposal to General Trudeau: "we could farm out to industry the components that comprised the electromagnetic antigravity drive and brain-wave directed navigational controls ... dole them out piecemeal once we broke them down into developmental units, each of which could have its own engineering track."

A list of the technologies which were eventually developed out of the Roswell back-engineering projects include: the silicon computer chip, "Star Wars" particle-beam weapons, portable atomic weapons, "invisible" stealth technology, lasers, integrated circuits, fiber optics, supertenacity fibers, night vision goggles, HAARP, and more. As Corso writes, no one before 1947 foresaw the invention of a technology based on silicon rather than carbon. The invention of the transistor and its natural evolution to the silicon chip was way beyond a "quantum leap." The development of the radio tube to the vacuum tube took fifty years, while the silicon transistor came upon us in a matter of months, writes Corso.

The transistor, which is the key component of the microcomputer industry, did not exist prior to 1947. Corso maintains that the transistor was chemically and functionally way ahead of its time when it "miraculously appeared during a two month window immediately after the Roswell crash." Its composition of silicon and traces of arsenic, for use as a superconductor, was completely novel. As Corso writes: "the means to identify what a transistor was, how it functioned, and how to fabricate one were beyond the technology of the time." He notes that even the chemical composition of the silicon had to be fabricated. Since it could not be told that this device was "reverse engineered" from found alien technology, a series of press releases created a two year history of the R&D of the transistor at Bell Labs. The cover story itself was reverse engineered in a feat of Orwellian Newspeak. According to Corso, the only person who would have known the true story was Bell Labs executive John Morton, who was murdered in the early 1970s.

The Roswell file included a contraption described as "tiny, clear, single-filament, flexible glasslike wires twisted together through a kind of gray harness." When held up to the light, Corso reports, an eerie glow emanated from this harness of narrow filaments, diffusing the light into different colors. Corso claims that, within five months of receiving this artifact from the Roswell crash, Bell Labs announced its discovery of the transistor. This fantastic story is also buttressed by the claims of Newark-based American Computer Company. The official history of the transistor holds that the first one was created at Bell Laboratories in Murray Hill, New Jersey, out of germanium, crystal, a few slivers of gold and a paper clip. The culmination of this research, led by John Bardeen, William Shockley and Walter Brattain, ushered in the Age of Information. The current incarnation of Bell Labs, Lucent Technologies, owned by John Lehman, is now celebrating the 50th anniversary of the first transistor.

Corso claims, had he not seen the silicon wafers from the Roswell crash with his own eyes, and heard reports of secret meetings at Bell Labs, he would have thought the invention of the transistor was a miracle. To back up this information, American Computer Company President, Jack Shulman, now claims to have mysteriously acquired a certain "shopkeepers notebook," dating from 1947, which previously belonged to Bell Labs. The information in this notebook suggests that "the transistor was back-engineered from an existing device." Shulman's story agrees with Corso in that the "history" of the invention was apparently given an Orwellian media spin. (*http://compamerica.com/roswell.htm*)

Shulman claims he has thousands of pages of supporting material which dates from this era. American Computer recently announced that it has developed a revolutionary device from information obtained from the same lab notebook which has the "potential to push computer technology ahead by a century." The device, which is called a "transcapacitor," is capable of storing up to 90 gigabytes of information while consuming very little energy. American Computer is predicting that it will manufacture a 90-gigabyte hard drive from this transcapacitor by late 1999 or 2000. While this story could likely be a hoax with stock market subsidies in mind, Shulman claims that he will selectively leak technology secrets "at the rate of one bombshell per month" in order to provoke government UFO disclosure. It will be interesting to see if anything comes of this.

With regard to the bodies of the EBEs, the Roswell alien autopsy reports that Corso studied indicated that the organs, bones and skin of the EBEs were different than that of humans. Corso claims that the heart and lungs were bigger, especially given its height of 4 feet and its small body build. The autopsy personnel at Walter Reed had noted that the brain was larger "but not at all unlike ours." The brain of the EBEs had four distinct sections and was oversized in comparison with its tiny stature; its cranium was more like palatal cartilage than hard bone. In addition, the blood and lymphatic systems seemed to be combined. Curiously, there were no food preparation facilities nor any stored food on board the craft; neither were there any waste facilities. Neither was there was a digestive or excretory system present in this biological entity. The autopsists noted that if any exchange of nutrients and waste took place, it had to have been through the entity's skin.

According to Corso, the skin of the EBEs consisted of a thin layer of permeable fatty tissue through which chemicals were likely exchanged with the combination blood/lymph system. The tiny mouth and lack of obvious digestive system was puzzling to the medical examiners at Walter Reed. They hypothesized that the EBEs recirculated waste chemicals, since there were no waste-processing facilities on board the craft. While reading the skin analysis report, Corso also speculated that the skin was "more like a house plant than the skin of a human being."

Another suggestion made by engineers at Wright Field was that there was no need for food preparation on this craft because it was a "scout ship" which did not routinely travel far from the mothership. The scout ship idea has been proposed by various researchers. The huge "mother" ships perhaps pass through interdimensional time/space portals for deep space travel, and the smaller scout ships which are held inside the larger ride the electromagnetic grid of the planets which they visit.

Corso suggests another possible explanation for the missing food and waste: they weren't actual life forms, only a kind of robot or android. According to Zecharia Sitchin in *Divine Encounters*, this idea is cross-referenced in the Sumerian account of a class of created emissaries called Gallu. The Gallu were described as not being made of flesh and blood. It is written in the Sumerian texts that the Gallu "have no mother, have no father," and that they "know not food or water." These emissaries were known to "flutter in the skies over Earth like wardens."

Indeed, the Walter Reed pathologists had noted that the EBEs were remarkably well adapted for long distance travel, having a heart and lungs of enormous capacity. Furthermore, the medical examiners suggested that the creatures probably had a much slower metabolism than a human's, since the large heart would beat more slowly and the milky fluid which was its blood was pumped through a primitive and reduced-capacity circulatory system. The pathologists stated in their report that the EBE's heart functioned as a passive blood-storage facility and a pumping muscle, but it did not work in the same way as a four-chambered human heart. They concluded that the heart's internal muscles worked less hard than a human heart muscle. In short, the EBEs "biological clock," Corso suggests, allowed it to travel great distances in a shorter biological time.

Corso noted that the EBE's large lungs occupied a large percentage of the small chest cavity, and seemed to function somewhat like a scuba tank, releasing atmosphere slowly into the system. In addition, the bones of the EBE were flexible and resilient as though, Corso speculated, they functioned as "shock absorbers," making their bodies well suited for physical trauma and extreme gravitational forces. The medical examiners concluded that the creature did not have to work very hard to sustain itself, and was especially adapted to long distance travel.

The bones and skin of the EBE showed a different atomic alignment, as if built for "greater tensile strength." Corso concluded that the skin may have been stronger in order to "protect the vital organs from cosmic rays or wave action or gravitational forces." Corso wrote in his report to General Trudeau that perhaps the EBEs represent "the end process of genetic engineering designed to adapt them to long space voyages within an electromagnetic wave environment, at speeds which create the physical conditions described by Einstein's General Theory of Relativity." (Corso, 98)

Brain Wave Guidance System

In short, Corso concluded that perhaps the EBEs were humanoid robots which were engineered for long distance space-time travel. In addition, he suspected that their technology made them part of the ship both with respect to piloting and navigation, as well as protection, by putting them essentially "in the eye of the hurricane."

In addition to the "inner skin" of the EBEs, the entities were enclosed in a one-piece protective jumpsuit which was worn like "an outer skin." When closely examined at Walter Reed, the medical examiner noted that this material reminded him of a spider's web, in that it appeared very fragile but was, in fact, extremely strong as well as flexible. This space suit appeared to be stretched around the EBE "as if it were literally spun over the creature and seized up around it, providing a perfect skin-tight protective fit." Medical examiners noted that the suit might have protected the wearer against low energy cosmic rays which regularly bombard the space craft during a space journey. By accelerating the atomic structure and heating the body, such particle bombardment might produce the effects of being cooked in a microwave oven. Without this material, suggests Corso, the oversized organs of the EBE would be vulnerable to the "cumulative trauma from constant energy particle bombardment." (Corso, 104)

Corso claims that the strands of lightweight fiber found in the Roswell crash reminded him of the filaments in a spider web, which exhibit natural supertenacity properties. He claims that this material, which was something like a ceramic/protein, had both encased the ship and formed the outer skin layer of the bodies of the EBEs. Tests on this fiber indicated that the molecules were very compressed and they were aligned lengthwise. A single strand could be stretched and twisted and convoluted, but would still return to its original length and tenacity. Corso realized when he saw this fiber that the key to producing it was to synthesize protein perhaps by replicating the chemistry of the spider gene, as well as by finding some method to replicate the lengthwise extrusion process of a spider. Corso was "fascinated by the prospect that something similar to a web spinner had spun the strands of supertenacity fabric around the spaceship." He speculated that if we could come up with such a material, it would give our own aircraft the same protection.

As it turned out, research into compressed molecular structures was already underway, and the Roswell material provided the impetus to continue work in this direction. The search for a molecularly aligned composite ceramic, he writes, led to the Stealth technology and depleted uranium invisible artillery shells successfully utilized in the Gulf War.

But there was more to this material than simply protection. The secret to the navigation of the flying saucer, Corso speculated, also lay in the skin-tight material which was seemingly "spun around the creatures," as well as around the craft itself. Somehow, he suggested to General Trudeau, the pilots "became part of the electrical storage and generation of the craft itself." The Air Force had discovered during its testing of the craft at Norton Air Force Base that the entire vehicle functioned somewhat like a giant capacitor. The craft itself, they wrote, stored the energy necessary to propagate the magnetic wave that elevated it, and allowed it to achieve escape velocity from the Earth's gravity. This internal energy also enabled it to achieve speeds of over 7,000 miles per hour. Corso described the vehicle as, essentially, a flying battery.

As suggested earlier, the smaller craft ride the magnetic grid of the planet, propelled by an electromagnetic force generated by rotating an extremely high-voltage electric charge around the hull of the craft. Corso takes it a step further to speculate that the vehicle essentially became an extension of the EBE's bodies by being tied to their neurological systems via their "space suits." This was how, he proposed, the EBE's became the primary circuit in control of the magnetic wave; in essence, "one with the vehicle and literally part of the wave." The pilots were not affected by the tremendous g-forces because they were, in effect, travelling "in the eye of a hurricane."

Also among the artifacts retrieved from the Roswell crash, according to Corso, were headband devices of flexible plastic material which contained electrical conductors. Officers at Walker Field in Roswell had even tried them on. Those whose heads were large enough to make full contact with all of the conductors got the shocks of their lives. Those who tried to rotate the device to touch different areas of their skulls reported various impressions, ranging from a low tingling sensation, searing head pain, or dancing or exploding colors inside their eyelids. Corso also connected this device to the piloting of the craft. These two technologies essentially comprised the electromagnetic antigravity drive and brain-wave directed navigational controls which the U.S. Army eventually farmed out to industry under the guise of "foreign technology" for purposes of back-engineering.

According to Corso, witnesses on the scene of the Roswell crash described receiving telepathic impressions from one dying creature. Yet, no one heard the EBE make a sound. Witnesses claimed they heard no words in their mind, but "only the resonance of a shared or projected impression" wherein they were "able to share with the creature a sense of suffering and profound sadness." But, since ultrasound scanning did not exist in 1947, there was no way to evaluate the nature of the cranial lobes. The examiners speculated about the psychokinetic and thought-projection powers of the EBEs, but could not offer any scientific data with regard to seeming telepathic powers exhibited by the dying alien. (Corso, 106)

The book *MILABS* explores the idea that the cloning of humans having a computer-brain has been proposed as a way to explore deep space. In a German television report entitled "Future Fantastic," televised on January 18, 1998, *X-Files* star Gillian Anderson gave viewers a tour of the possible future of cybernetics. The creation of bio-engineered robots, or "cyborgs," the television show explained, would necessitate the cloning of humans who have computers instead of brains. In a proposal to NASA, cyberneticist Dr. Manfred Clynes has suggested the concept of re-engineering humans to operate in deep space without space suits. The cyborg astronauts would be sustained by intravenous pumps and slow release drug systems. According to Martin Caiden, certain military personnel volunteered their own bodies to science and became cyborgs in 1970. *MILABS* also reports "rumors" that the Air Force is conducting cybernetics projects to modify humans for space flight. Could this be what is going on in underground military bases, where fetuses in tubes have been witnessed by abductees? Is this yet another back-engineering project which came out of the Roswell crash? Corso does not mention this one. Of course, once these artifacts were out of his hands and into those of the CIA/Nazi/occult underbelly, we can only speculate as to what devilish ideas were later spawned.

The Virtual Merkaba

There are some remarkable corroborations between Dr. Jacobs' research into the outer morphology of the alien grays and the deductions made by Philip Corso from his analysis of the artifacts and reports contained in his Roswell

File. According to the medical examiners at Walter Reed, this entity did not have to work very hard to sustain itself, and was especially adapted to long distance travel. Corso takes it a step further. He surmises that the entity is "part of the electrical storage and generation of the craft itself" via its own neurological system, as well as by the addition of several different technologies. The bizarre headband device and the specially 'shrink-wrapped' outer skin completes the package. This astronaut is ready for takeoff.

In turn, via a different analysis, David Jacobs has deduced from abductee descriptions of the morphology of the alien grays that the alien environment is "highly controlled, uncontaminated, unnatural and highly technical." Thus, we can safely deduce that the highly technical environment which the grays call 'home' is none other than the aircraft itself. They are *engineered* as one unit. It is important to note that, according to Corso, the reports from the Roswell crash indicate that these beings did not have a chance of survival outside of their craft. The androids and their ship are a package deal. Since living beings cannot travel through time and space due to energy particle bombardment as well as other effects, this technological design is quite possibly the *only way to fly*.

But this means that someone engineered them; someone came up with this design. It would appear that this high-tech package is somehow connected with the idea of the universal constructor or messenger probe. Yet, as the Space Travel Argument contends, von Neumann probes are *closely analogous to bio - logical species*. In his book *Nothing in This Book is True, But It's Exactly How Things Are*, Bob Frissell talks about the *merkaba*: an internal time-space vehicle which can only be achieved by advanced meditative states. The internal merkaba is not functioning on most people, but it is used by ascended masters to travel at the speed of thought.

The merkaba is a time-space vehicle created by counter-rotating fields of electromagnetic energy which can only be created with the emotional body intact. Frissell explains: "The mental star tetrahedron, electrical in nature and male, rotates to the left. The emotional star tetrahedron, magnetic in nature and female, rotates to the right. It is the linking together of the mind, heart and physical body in a specific geometrical ratio and at a critical speed that produces the merkaba."

According to Frissell, the word *mer* denotes counter-rotating fields of light, *ka* is spirit and *ba* is body or reality. The merkaba is made up of counter-rotating fields of light which encompass spirit and body. It is, according to Frissell, "the image through which all things were created, a geometrical set of patterns surrounding our bodies." The geometry of the merkaba surrounding the body and spirit forms all at once a star tetrahedron, a cube, a sphere and an interlocked pyramid.

According to Frissell, the "vehicle of the deity," the merkaba, can be externalized. All flying saucers, he notes, are based on the principles of the external merkaba. To manufacture an external merkaba, nothing from the emotional body is needed; it is all *mind*. Creating an external merkaba, however, turns emotions into technology. Therein lies its fatal flaw. The construction of ships to cross the stars and robots to do all the work, Frissell warns, robs species of their emotional bodies, which is their main source of power. He writes: "The grays are a race without an emotional body and they are dying because of the fatal flaw associated with basing creation on the external merkaba." The creation of the external merkaba, Frissell posits, "traps them in their present level of existence."

According to ancient Vedic texts, many humanoid races are said to possess *siddhis*, which are powers claimed to be based upon "natural principles."

Richard Thompson writes in his book *Alien Identities* that it is possible that machines might be constructed which would take advantage of these natural principles. The *siddhis* include: the power to change the size of objects or living bodies, the power to move from one place to another without traversing the space between, the power to travel through physical objects, the power of long distance hypnotic thought-control, invisibility and cloaking, the power to assume various forms or generate illusory forms, and more.

Are flying saucers and the androids inside them essentially based on the idea of the external merkaba: counter-rotating fields of electromagnetic energy, in a specific geometrical ratio and at a critical speed, encompassing body and spirit as one unit? Are they also based on the *siddhis*, as natural powers existent in the Universe? As Corso has noted, the ship stores the energy necessary to propagate the *magnetic* wave that elevates it, and the androids inside become part of the *electrical storage and generation* of the craft itself. As Corso has noted, the ship is essentially an extension of their bodies, tied to their neurological systems via their techno-gear, making them "the primary circuit in control of the magnetic wave." Indeed, the rotating fields of light surrounding the body—male energy spinning left, female energy spinning right—might very well be described as "the eye of the hurricane." This leads to another interesting connection: that the androids are androgynous, neither male nor female, but something else. Or, as Sitchin translates it, the amorphous creatures are not asexual but are *both* male and female: trans-sexual.

Smart Probes and Indefinite Survival

A messenger probe is essentially an advanced computer sent to another stellar system, which includes an intelligent universal constructor capable of self-repair, self-programming and self-replication. The universal constructor receives instructions from its system of origin to carry out a search for interstellar construction materials in order to build a colony or base in the stellar system. The probe is capable of building smaller probes to move around within the stellar system. So, then, the job of a universal constructor is to construct.

The Space Travel Argument contends that the launching of a universal constructor would increase the probability of the survival of a civilization in the event of the death of its star, nuclear war, or other catastrophe. In order to accomplish "indefinite survival," space colonization would not need to be of an 'imperialist' nature, but could consist of the construction of space stations revolving around stars. They explicitly state "those groups which do not exhibit this behavior would be *selected against*." So, in Martha Stewart's words, 'it's a good thing.'

The Space Travel Argument contends that if a civilization possessed artificial womb technology, they could program a probe to synthesize members of their species in the other stellar system. These members could then be raised in something like an O'Neill Colony by robots also manufactured by the probe. The Space Travel Argument also presumes that eventually such beings would be free to develop their own civilization in the other stellar system. The Space Travel Argument also seems to put forth the idea that it is wise to populate the stars with "machine descendants" since that civilization would never become extinct. If we accept the assertions of the Space Travel Argument, robots or androids should be recognized as intelligent fellow beings and accorded full human rights.

The probe is essentially a collector of information, and the information it collects are the building blocks of life. If it is discovered that there is indeed biological life in the stellar system to which a probe is sent, the civilization could also consider those life forms as a source for "construction materials." They might use the building materials (DNA) of those life forms to enliven its possibly degenerating DNA, to create a race to populate a local colony, or to send to other star systems for distant population projects. Thus, if there were biological life forms in the system, they would not need "artificial womb technology." They could utilize real wombs and harvest the fetus at such a time that their technology could take over, or they could abduct citizens of the planet to use as temporary breeders.

As Corso contends, the EBEs "are fabricated beings, just like robots or androids, created specifically for space travel and the performance of specific tasks on the planets they visited." He speculates that "perhaps their programming could be updated or altered from a remote source." (Corso, 245) As engineers at Wright Field suggested, perhaps the smaller craft which crashed in New Mexico was a "scout ship" which only travels locally and does not venture far from the larger craft, or from a colony in outer space. Perhaps in our thinking of them as "ships" or "saucers," or simply flying machines, we have overlooked a more accurate description and, therefore, have missed the true meaning of what they are: the computerized "smart probes" of an extraterrestrial civilization.

If we view the abducting gray entities and their UFOs as one unit, and as the programmed "universal constructors" of a vastly superior race, we have an explanation for the procreative nature of the abduction scenario. In considering the origin of human civilization on Earth, Robert Temple suggests in *The Sirius Mystery* that we should now be ready to face the reality that extraterrestrial civilizations exist and have been here. He surmises that if intelligent beings have already visited Earth, they may possibly be "monitoring us at this moment with a robot probe somewhere in our solar system, and may have the intention of returning in person some day to see how the civilization they established is really getting on."

Has it now become fully obvious to all but the most hardened skeptics that the grays are essentially traveling biochemists, collecting the DNA of biological life forms on Earth? Does it make any sense at all that the flying saucer and the android within constitute the probes, or universal constructors, of an advanced spacefaring civilization? Does it not make sense that an advanced civilization, especially a non-humanoid race, would utilize humanoid robots to collect information from a humanoid race? The "humanoid interface" theory will be addressed in detail shortly. Let's first address some of the possible reasons for the stepped-up "sampling" schedule since 1947.

Virus X and the Cell From Hell

Many UFO researchers have noted the obvious correlation between the first atomic bomb explosions in the southwestern desert area of the United States and the first sightings of UFOs in 1947. Recall the ancient *Virgin of the World* treatise which suggests that "mankind have been a troublesome lot requiring scrutiny and, at rare intervals of crisis, intervention." The babysitting of the human race is obviously on red alert status due to the concerns that the children are playing with fire. The Earth environment has become unstable due to Oppenheimer's mistake, and it threatens the entire solar system, and perhaps even beyond. It also threatens the priceless DNA structure of the life forms on this sacred blue planet.

In addition, new viruses and virulent new strains of old viruses are now emerging at a fast clip. In his book, *Virus X*, Dr. Ryan explains why so many plagues are currently ravaging the world: a huge population explosion (6 billion people alive today versus 1.5 billion a century ago) together with the destruction of many habitats has brought human beings into contact with viruses that once lived beyond our reach. Killer viruses that have been in the news of late have names like Ebola, Lassa, Marburg, Hantavirus (Navajo plague), AIDS, HIV, Gulf War syndrome, Mad Cow Disease, Pfiesteria piscicida, Rift Valley fever (Egypt), Junin virus (Argentina), Necrotizing Fasciitis, Tobacco Mosaic Virus, Epstein-Barr, Hepatitis, Herpes, Epstein-Barr virus (perhaps the first known human cancer virus), Crimean-Congo hemorrhagic fever, the re-emerging Smallpox and Polio viruses, and good old fashioned Influenza.

What makes many of these bacteria even more resilient is the effusion of bacteria-killing products on the market, along with an attitude of fear that goes hand-in-hand with the use of these products. This attitude is summed up by the following quote found recently in a "beauty" magazine: "Everyday things continue to become more glamorous. The kitchen-sink hand wash from Crabtree & Evelyn is pale-green, smells faintly of cucumbers (a scent that lingers pleasantly on your hands after they've been washed), and kills bacteria (*Salmonella*, *E. coli* and mad-cow disease have replaced ax murderers in my imagination)."

Until just a few years ago, horrible viruses such as Ebola, Hantavirus, and Marburg were as unknown to most of the world as the places in which they originated. Pfiesteria piscicida, the "cell from hell," an aquatic microorganism found in the polluted waters of North Carolina, causes symptoms similar to Alzheimer's or multiple sclerosis. In his book, *Virus X,* Dr. Ryan examines the recent work of microbiologists working on the causes of AIDS, Ebola, and the Navajo plague. Ryan proposes that somewhere in the undisturbed parts of the rain forest lurks a virus (Virus X) that could annihilate humanity via its "genome intelligence," an outgrowth of high mutation rates.

While plagues and pestilence have always been part of human existence, humans tend to acknowledge these primitive co-inhabitants only when they are mowing down entire populations, like the AIDS virus is now doing on the continent of Africa. *The New York Times* reported on 8/6/98, Zimbabwe has the highest AIDS infection rate on Earth, with a stunning 90% of the world's AIDS deaths. Four million people were newly infected with HIV in sub-Saharan Africa in 1997. AIDS has already killed more than ten million people in that region alone, and it is projected that twenty million more will die in the next few years. Zimbabwe once had a health system that was "the envy of southern Africa." Now, with an average of less than $10 to spend on each person's health every year, most African countries cannot fight the big fight. Many will continue to die of AIDS, as well as malaria, hepatitis, tuberculosis, measles and cholera.

Humankind: The Sixth Extinction

Earth ecosystems are being strained by overpopulation, with approximately 27,000 species vanishing each year as a result of human activity. In addition, 16 million children a year die from hunger and malnutrition on this planet. With biological warfare being threatened all over the planet by both human and microbial cohorts, and complacency and denial at an all-time high, it is no exaggeration that the human race itself is also being threatened with extinction. On five occasions in the past, catastrophic natural events have caused mass extinctions on Earth. In *The Sixth Extinction*, Richard and Roger Lewin consider how the human species is wreaking havoc around the world. Considering the past history of the five great extinctions which have taken place on the planet, these authors argue that the sixth great extinction pattern, spawned by mankind itself, has already begun. Humans alone are in the unique position to trigger the sixth extinction in the history of life. They are also in the unique position to do something about it, but that's another, happier story.

The Permian extinction wiped out 96 percent of all species on Earth, as well as perhaps 99 percent of the individuals comprising the surviving 4 percent of the Earth's populations. In his book, *Mobius*, (*www.protect-Earth.com*) Richard Hofstetter reminds us that the one species that survived, and always survives, is Gaia. She is at the top of the food chain. "Nothing lives as long as Gaia," he reminds us, for "her life span is measured not in terms of seasons or decades, but in eons."

In this bizarre mix of sociobiology and Gaian animism, Hofstetter suggests it is Gaia's job to make the world comfortable for other living creatures in order to assure her *own* survival and longevity. In this task, she seems to strive for more and more biodiversity. More than 99% of all species ever to inhabit the Earth are now extinct. Extinction is a way of life: species come and go, but the Earth endures. Hofstetter claims that humans have become so numerous and so aggressive that we threaten to trigger the extinction of numerous Earth species. We are also a threat to the Earth's climate. Hofstetter bets that Gaia will survive these "human perturbations," but humans may not. He claims the real task of environmentalists is to preserve the "conditions most conducive to the survival of *Homo sapiens*," and that we can do this by "preserving biodiversity and by leaving smaller footprints."

Gaia, no different from any other living creature, is interested in her own survival. She sheds species, Hofstetter suggests, "the way we shed epidermis." Hofstetter believes that humans are in the midst of a great evolutionary leap, which has made us "the enantiomorph of this ancient and distinguished life form." He disagrees with environmentalists who believe that humans are solely a destructive constituent of the Earth's environment. Claiming that humans are part of the Earth's physiology, he suggests that, as the evolving Stewards of the Earth, we must become partners with the Earth in a more efficient enterprise. He sees much that is good in humanity, and sees humans as Gaia's "chosen species," whom she may even forgive "for trampling a few daisies as we gather our wits."

Hofstetter's banal understatements may be fashionable at green cocktail parties, but his macho sociobiology married with Gaian fem-animism creates a rather queer juxtaposition. He correctly notes that "human behavior maximizes the inclusive fitness of the *individual*, not of the species," while flippantly suggesting that we have some sort of understanding with Mom that we'll mop up our play room later. Humans, especially technological civilizations, have trampled tons of *daisies*, and our *wits* are still nowhere in sight. The Earth will forgive us for nothing. She'll throw us off her back in a New York minute. As a matter of fact, it appears that Gaia routinely does a performance art dance which effectively douses the fleas on her back with a constant bombardment of killer viruses.

There have always been plagues on Earth, but the newest designer plagues are, ironically, being caused by the technology which puts us "in touch" and makes the Earth "smaller." We have already trampled more than our share of daisies, and in doing so, we have let loose a horde of virulent plagues and pestilence we never realized was possible.

Hofstetter is too busy making excuses for Gaia's favorite pet to realize that microbes also play a role in creating and maintaining the biodiversity that Mother Earth loves. The major purpose of plagues and epidemics is not population control. Rather, the germs that cause them are merely trying to stay alive. Contrary to this ridiculous Stewardship model, the reality is that humans are not Gaia's "chosen species" any more than any other. She apparently loves us all the same, even the lowly, microscopic "intelligent germ." In his book, *Yellow Fever, Black Goddess: The Coevolution of People and Plagues,* Christopher Wills puts such exotic life-forms on center stage. The germ has a right to live too, even if they appear to exist at the "very edge of the possible." They are simply maximizing their potential in the race for indefinite survival. And it looks like they might win the War for This World.

Human Embryo Cloning: The Other Sex Scandal

To add insult to injury, the human genome is degenerating. Research has shown that males born after 1970 have less than half the sperm count of males born prior to 1950, with a much higher incidence of genetic abnormality. Some biologists have suggested that it may become necessary to alter the human genome in order to repair ongoing genetic damage which is causing a marked increase in genetic abnormalities over the past century. Most scientists believe these abnormalities are the result of environmental damage caused by the excessive use of pesticides and chemical contamination of water. Today, a person is four times as likely than they were just a hundred years ago to be born with a genetic disease. This is why human embryo cloning is the top news story today, buried under presidential sex scandals. Embryo cloning is a technique used by researchers and animal breeders to split a single embryo into two or more embryos containing duplicate genetic information. The procedures used in embryo cloning have been around for many years, and have been used in the cloning of cattle and sheep embryos for the production of animals with known genetic traits.

Human embryo cloning is the Other Sex Scandal of 1998. During the 1980s and 1990s, research in human embryo cloning was banned by both the Reagan and Bush administrations due to pressure from pro-life factions of the Republican party. However, the pro-choice Clinton administration has a somewhat different attitude, toward both sex and human embryo cloning, and has relaxed the regulations against the federal funding of fetal and embryonic research. Federal guidelines for human embryo cloning have been established. Research is allowed only on pre-existing embryos, with development allowable up to and including the fourteenth day. New embryos are allowed for only what the National Institutes of Health considers "compelling research." Many other controversial areas of potential research have as yet been undecided, but research into human cloning is simply moving to countries which don't have laws against it. (see *http://cac.psu.edu/*)

As a result of relaxed mores, Robert J. Stillman of George Washington Medical Center in Washington DC announced in 1994 that he had successfully cloned 17 flawed human embryos. On January 6, 1998, Richard Seed informed NPR's "All Things Considered" that he and his colleagues are set to begin work on human cloning this year. NPR's science correspondent, Joe Palca, reported that Seed and some Chicago-area doctors have the skills and equipment to attempt the cloning procedure on humans. It appears that certain groups may proceed with human cloning in countries where it is not illegal, since its ultimate purpose may be for the genetic re-engineering of the human species due to non-repairable genetic damage: the winding down of the genetic clock of humankind.

Ironically, the *Washington Post* ran a story on May 27, 1999 indicating that molecular studies conducted on the sheep "Dolly," which was cloned in Scotland, have disclosed that clones inherit the age of the animals from which they are produced. It is predicted that, as a result of inheriting her progenitor's six years of genetic wear and tear, Dolly will die much sooner than was expected. This new development puts an interesting spin on things. In our futile attempts to use science to turn the clocks back, we find we can only turn them forward.

A Bahamas-based company, Valiant Venture Ltd, will soon be offering a service called "CLONAID" to provide assistance to persons who wish to have a child cloned at a cost of approximately $200,000. CLONAID also offers a service called "Insuraclone" which provides for the sampling and safe storage of cells from a living person in order to create a clone after death. CLONAID is the first company in the world to offer human cloning, and expects to have over a million customers interested in its services and as many laboratories interested in working partnerships with them. Ironically, the company was started by the Raelian movement, a religious organization which claims that life on Earth was created in extraterrestrial laboratories. (*www.clonaid.com*)

If at any time in Earth's history this planet had been visited by an ETI civilization which has at least "noted" its human presence, the preservation of the genetic heritage of this planet would be the prime motivation of the increase in "sampling" behavior.

"Star Wars" Defense and the Gray Menace

In his book *The Day After Roswell*, Col. Corso claims that stealth technology was developed by back-engineering the alien craft which crashed in the desert near Roswell in 1947. The U.S. Army, he claims, was mainly interested in how the weapons system could be re-engineered for defense use. The Air Force, however, was more interested in adapting the propulsion and navigation systems to our level of technology. Back-engineering experiments took place at Norton Air Force Base where, Corso claims, the Air Force and the CIA still maintain a sort of "alien technology museum." Experiments continue to this day, Corso claims, almost in plain sight for people with security clearance. Those who are shown the secrets are immediately bound by national security laws. The 'crash test dummies' media spin shows that the Air Force is still working to keep the truth of this matter under wraps.

Corso's detailed descriptions of the Roswell craft indicate that there was nothing conventional about the Roswell craft's propulsion system. During the first few years of testing at Norton, the Air Force discovered that the Roswell craft functioned like a giant capacitor. The craft worked by displacing gravity through the propagation of magnetic waves controlled by shifting the magnetic poles around the craft to repulse "like-charges." Defense contractors raced to figure out how the ship retained electric capacity, as well as how the pilots inside could live within the energy field of a wave. This race was all about who was going to land "multi-billion-dollar development contracts for a whole generation of military air and undersea craft."

According to Corso, however, it wasn't just about money. Corso contends that there was, in fact, a real threat from outer space. It was eventually discovered that the technology of the ETs was along similar lines of Tesla technology: wireless transmission of energy, and manipulation of gravitational field through electromagnetic wave propagation. In addition, Tesla's "death ray" became the basis for the defense weapons which, Corso claims, were deployed to challenge "hostile intrusions" of our airspace by the ETs. These "two confluent streams of scientific theory," the back-engineering of the Roswell craft at Norton combined with Tesla-based technology, eventually became the basis of President Reagan's Strategic Defense Initiative (SDI), an antiballistic missile defense system. Corso claims that particle-beam weapons and "Star Wars" defensive weapons, put us on "a wartime footing with the EBEs."

Corso writes that the first type of weapon which was utilized against "space vehicles" entering Earth's atmosphere from space was an accelerated particle beam weapon. A workable directed-energy particle beam weapon was still under development as part of the SDI. Corso writes that the Advanced Research Projects Agency (ARPA) had a clue that the only possible weapon that could interfere with the electronic field drive of the EBEs aircraft was the directed-particle energy-beam weapon, which worked by disrupting the electromagnetic wave formation around the spacecraft and penetrating the antigravity field. But incoming Soviet missile defense was in some ways a cover story for the true necessity for the ultimate in "Star Wars" defense. In a conventional sense, the particle-beam weapon in deployment as part of the "Star Wars" antimissile defense system worked by destroying electronic systems with its directed electromagnetic pulse. It also excited the atoms in the target, causing it to explode. But in the case of more "unconventional" use against non-terrestrial unfriendly visitors, the particle beam worked in a different way. The powerful electromagnetic pulse would disrupt the antigravity field of the vehicle, forcing the vehicle to crash by "destroying its ability to counter gravity."

Corso claims that the SDI changed the Cold War by showing the Soviets we had a real nuclear deterrent. He asserts that this technology, at the same time, "forced the extraterrestrials to change their strategies for this planet." Through "limited deployment of the SDI," Corso writes, "humanity won its first victory against a more powerful and technologically superior enemy who discovered, to whatever version of shock it experiences, that there was real trouble down on its farm." Corso's use of the word "farm" is almost as interesting as his perception that any victory has been won over the EBEs at all.

The Politics of Flying Saucers

By establishing its sham bodies of official investigation into UFOs (Project Sign established on January, 22, 1948, Project Grudge on February 11, 1949, a "new" Grudge in October of 1951 and Project Blue Book in March, 1952) the U.S. government was able to convince people that they were dealing with the problem of the invasion of Earth skies by intelligent life from somewhere out there. Once the classified files of Project Blue Book were opened, long after its demise, it was obvious that it had for two decades served as a cover to make the American citizens think their chosen representatives were actually looking into the subject. In fact, they were! The trick is they formed overt investigative bodies whose function was to come up with nothing, while covert groups, which remain 'classified' to this day, really delved into the matter. If UFOs do not exist, why did the government form a secret committee called Majestic Twelve?

As more citizens decried the government's bogus antics, its response was to set up more bogus "studies" and "investigations." This has continued up to the present. In 1995, Congressman Schiff of Texas requested a report of the Roswell incident by the U.S. General Accounting Office Report No. GAO/NSIAD-95-187. A paper published by Project VISIT in August 1995, entitled "Project VISIT Reviews the Results of a Search for Records Concerning the 1947 Crash Near Roswell, New Mexico" concluded that "the GAO did not go and inspect the Roswell records themselves and they did not hold any meetings with other agencies to define search protocols..." and that "only a cursory search for records was done." The entry for the period of July 1947 quotes the official answer: "a radar tracking balloon."

The two pages of the Roswell Army Air Field Combined History for July 1947 does not provide an answer for the Roswell incident. Instead, states John Schuessler of VISIT, "it documents the lie told in the first place." VISIT's focus is the engineering and study of the internal systems of the spacecraft of visitors, with a secondary interest in the beings inside them. A published document elucidates the cover-up by the FBI, CIA, NSC, DOE, DOD and other alphabet soup agencies. [SASE and $1 for copying costs to VISIT—Vehicle Internal Systems Investigative Team, POB 890327, Houston, TX 77289] In addition, the GAO audit of the paper trail surrounding the Roswell crash, "indicates that many important records, such as the outgoing messages from Roswell Army Air Field, were destroyed without proper authority."

According to Col. Corso, numerous secret projects were started since the Roswell crash in order to manage the ongoing extraterrestrial problem. At each level, he discloses, once security was breached for any reason, "even by design," a part of the secret would be declassified and the rest would be brought under the wing of a new classified project, or moved under the umbrella of an existing one. He explains, the government is not "some monolithic piece of granite that never moves or reacts," but is highly proactive, especially when it comes to protecting its secrets. Since the Roswell affair, Corso writes, the government was "a hundred steps ahead, a thousand, or even more." The government never hid the truth, he observes, they just "camouflaged" it. It was always there for people to figure it out for themselves. And, he claims, they found the truth, "over and over again."

For instance, Corso explains that Project Blue Book was created to give the people a place to report saucer sightings, and Projects Grudge and Sign were created for sightings and encounters that couldn't be explained away. Corso also outlines secret programs like the planned Project Horizon Moon base, the High Altitude Research Project (HARP), and the Strategic Defense Initiative (SDI), which all have something to do with alien technology and defense of the Earth.

Corso maintains that the EBEs were not benevolent space beings who had come to enlighten Earth humans. He claims they were "genetically altered humanoid automatons," or "cloned biological entities," who, he asserts, were harvesting biological specimens on Earth for their own experimentation. And, he admits, as long as we had no weapon against their superior technology, our government had no choice but to allow the harvesting of the human race by alien entities to continue. Curiously, he writes, "we had negotiated a kind of surrender with them as long as we couldn't fight them. They dictated the terms because they knew what we most feared was disclosure." But this changed in 1974 when the US Air Force used a particle-beam weapon to bring down an alien craft over Ramstein AFB in Germany. The EBEs knew then, Corso asserts, that our defense of the planet was in place.

It has been suspected that the initial reason for denial of the existence of interplanetary visitors may have been that the government/military was powerless to do anything about it. As a matter of fact, Corso agrees. Corso writes that there were hints during the Cold War era that there was a "hidden agenda" in operation, and there was heavy censorship in the military. Alien abductions were of grave concern to the UFO working group. If the government couldn't protect its citizens from alien abduction, it didn't look good. Likewise, if too many people began to see real ET spacecraft, it would become obvious that the government couldn't protect its citizens. For a time, Corso admits, this was true. Initially, the government couldn't protect its citizens. But with the deployment of "Star Wars" particle-beam weaponry and HAARP, this situation changed. In his book, Corso outlines the course of events which, *he believes*, forced the EBEs to back off and find another approach.

Corso claims that Star Wars technology forced the extraterrestrials to back off and change their normal way of approaching humanity. Blatant contact of the sort which seems to have occurred in Biblical times cannot happen now due to our acquired level of technology. To this degree, the Roswell crash changed technology and reality, and altered the course of history. In keeping with doctrines which ignore 'externality' to the greatest degree possible, part of the official denial of the Roswell crash rests with the pervasive attitude that things just don't 'drop in' from externalia. Therefore, it is ironic that a former high-ranking member of the military is now conceding that things do indeed drop in, and that this very paradoxical event is what gave humanity the ability, as least temporarily, to stave off hostile extraterrestrial threats.

Since it defines what it considers proper evidence, the Space Travel Argument seems to hold in place an entirely engineered reality system. The fundamental lie at the bottom of this thought-conforming design is Darwinian evolution. Isolationist, materialist, and localized cause-and-effect Western paradigms, particularly Darwinian evolution and the gradualist paradigm, as well as Freudian mechanisms, serve to keep ideas of our origins insulated and inbound, and keep us from possibly adding up 2+2 with regard to our true ancestry from the "sky" rather than the "water." Let's investigate whether this curious Western philosophy can really be considered an objective and scientific hypothesis, and why, regardless of its unscientific nature, it is peddled as the most ultimate truism.

Chapter Five

The Cart Before the Horse:

Darwin and the Origin of the Humanoid Form

The Universe has an obliging nature and is reflexive. It can provide proof for any cosmological scheme, scientific or mystical, foisted upon it.

Mann & Sutton

Giants of Gaia

From the perspective of Darwinian theory, mankind may be seen as the winner of a preposterous survival lottery which, we are told, given the incredible odds should not have occurred even once. Thus, the logical deduction from Darwinian theory is that if the humanoid form evolved from the great ape lineage on planet Earth, the mathematical odds are incredibly against the possibility of that same chain of random and incremental steps, contingent upon an interplay with a similar biological environment, occurring elsewhere in the Universe. Therefore, the assumptions of Darwinian evolution presuppose the humanoid form to be an entirely Earth-based phenomenon. Remarkably, the fundamental assumption of the accidental evolution of mankind from the great ape lineage is always overlooked as the problematical factor in the analysis.

An example of this common presumption is stated succinctly in an interview which appeared in *Paranoia* magazine's Winter 1997 issue. In D. Guide's interview with Henry Stevens of the German Research Project, Stevens' discussion of terrestrial-based flying saucer technology includes the comment that: "If a creature has two arms, two legs, walks bipedally and has stereoscopic vision, it is a human or a human derivative in my book. Parallel evolution would not produce such a close analog on another world."

Although arguments on the side of terrestrial-based UFO technology are certainly valid, it is not within the scope of this work to address the nature or extent of Earth governments' covert preoccupation with flying saucers, mind control or paranormal studies. One of the aims of this book is to debunk Darwinian evolution as a testable scientific hypothesis from which to argue mankind's singularity or uniqueness in the Universe.

The concept of "scientism" is the total commitment to a materialist worldview. As Charles Tart writes ("Six Studies of Out of Body Experience," *Journal of Near Death Studies*, 1997): "since scientism never recognizes itself as a belief system, but always thinks of itself as true science, the confusion is pernicious." Tart believes a scientist should be committed to observe things carefully and honestly, then devise theories and explanations about what those observations mean, without *ad hoc* rationalizations. Tart suggests that those who ritually practice "scientism" have an emotional attachment to a materialist worldview, a belief which in itself should be subjected to continual testing and modification. As Tart writes:

> If a theory has no empirical, testable consequences, it may be a philosophy or religion or personal belief, but it's not a scientific theory. Science has a built-in rule to help us overcome our normal tendency to become emotionally committed to our beliefs. This is where scientism corrupts the genuine scientific process.

According to the above definition, Darwinian evolution is "scientism." It is not a testable scientific hypothesis. It is an emotional commitment to a highly-touted underline philosophy of Western materialism, which has the major backing of Earth's reality engineers for reasons which *seem* apparent (financial and emotional investment), but which ultimately remain elusive. The following analysis will show that Darwinian evolution constitutes a tautology: a self-contained system of circular proofs, which are always true in a self-contained system of circular proofs. If it can be shown that Darwinian evolution is not a valid scientific theory, it follows that any argument following from it must be considered merely an extrapolation rather than a logical deduction. This book will argue that Darwinian evolution cannot be used as a framework from which to correctly argue against the Cosmic co-existence of the humanoid form, or human-like intelligence, since it likely places the cart before the horse.

The most common argument against the existence of 'intelligent'life in the Universe is based on the consensus-reality of Darwinian evolution. To state it more specifically, the fundamental premise underlying the argument against the existence of intelligent life in the Cosmos, and specifically the humanoid form, is the assumed impossibility of the separate evolution of upright, bipedal, large brained, tool-making hominids on planets which are worlds apart.

A confounded dilemma trips up the popular use of the evolution argument against the co-existence of the humanoid form in the Cosmos at large. As stated earlier, we are extrapolating this presumption from an unproven theory based on an Earth-centric bias. Astronomer Tom van Flandern has noted the erroneous assertion that the probability of ETI visiting our solar system is 'extremely small.' He notes that since this presumption is not a known scientific fact, the probability of ETI visitation is actually 'unknown.' After all, wouldn't the appearance of an ET race in our skies automatically make short work of Darwinian evolution?

The philosopher William James asserted that empiricism demands that we "look at a range of experience seriously and open-mindedly, and consider what is the best way to describe it, rather than defining it in advance in ways designed to outlaw alternative descriptions or forms of it which we find inconvenient." As logical empiricists with our minds and hearts open wide, and with no biases either way, let us now attempt a clear examination into Charles Darwin's theory of the natural selection and evolution of Earth species, and its extrapolation as a cosmic constant.

A Chain of Accidents

As an undergraduate anthropology major at a southwestern desert university, my first physical anthropology course was quite an experience. Although I accumulated an immense amount of knowledge that semester, it was the first meeting of the class that I will never forget. In the midst of jokes about "noses running in my family" and so forth, there was an unsettling undercurrent in the introductory dialogue. The instructor, a Ph.D., was not so jovial about one thing. With an angry and reddened face, she proclaimed that Darwinian evolution was a fact and not a theory, and warned us in no uncertain terms that she would entertain no questions with regard to the facticity of evolution. What struck me as odd at the time was her tone of exasperation at even the anticipation of an underling wasting her time arguing this *fact*.

Well, noses run in my family too. I knew, right off the proverbial bat wing, that something smelled fishy, but it took me several years to realize that she was only one of the countless college professors, biologists, science writers, scientific researchers, philosophers, and publishers with a vested psychological, emotional and financial interest in Darwinian evolution. Evolutionary theorists bank on the hope that this theory is too complicated for most of us to fathom, and that we will not ask questions out of fear of appearing ignorant of the *facts*. More often than not, however, the disconcerting questions which most people have about evolution are very appropriate and intelligent. The truth is, some logic and a little horse sense is really all you need to understand what Darwin was trying to say. It's the mess that his followers, so-called neo-Darwinists, have made of it that often takes real patience to decipher.

It is clear that the theory of evolution essentially views the human form as merely an accident in a chain of accidents. For instance, Stephen Jay Gould argues that the evolution of the human form is not a "repeatable occurrence." In the *Journal of British Interplanetary Society* (1992), E.J. Coffey argued that "the evolutionary pattern shows rapid diversification followed by decimation with perhaps as few as five percent surviving" and further that "the survivors resemble the winners of a lottery rather than creatures better designed than the unlucky majority who do not survive."

British astronomer Sir Fred Hoyle has mathematically dismissed the chance of evolution being an actual occurrence, arguing that "even if the whole Universe consisted of organic soup, ... the chance of producing merely the basic enzymes of life by random processes without intelligent direction would be about 1 over a 1 with 40,000 zeros after it; a probability too small to imagine." Hoyle concludes that "Darwinian evolution is most unlikely to get even one polypeptide sequence right, let alone the thousands on which living cells depend for survival." Given that there are trillions of different kinds of cells in the body, all in delicate balance with each other, each of these varied cellular structures would also have to develop by chance. In a *Times-Advocate* interview in December 1982, Hoyle declared that this mathematical impossibility is well known to scientists, yet nobody seems willing to "blow the whistle" on the absurdity of Darwinian theory. Hoyle claims that "most scientists still cling to Darwinism because of *its grip on the educational system*." They do not want to be "branded as heretics."

Taking the Super Out of Natural

The first assumption Charles Darwin made in his research into genetic variation between parent populations and their descendants was that species are not immutable but, rather, "descent with modification" is the norm within species. He proposed that this process of change can account for all, or nearly all, the diversity of life. He thought that it would, some day, be proven that all living things descended from a common ancestor, and perhaps even a single microscopic ancestor. Darwin proposed as a mechanism for this process a concept he called "natural selection." He later regretted use of the word "selection" since it seemed to give the concept a teleological air which only served as fuel for his rivals.

The National Academy of Sciences has told the Supreme Court that the most basic characteristic of science is "reliance upon naturalistic explanations" as opposed to "supernatural means inaccessible to human understanding." That's funny. Human beings have cultivated a comfortable relationship with things "supernatural" over the course of their days on Earth, while it might be said that the relatively newfound theory of Darwinian evolution has made itself very inaccessible to human understanding indeed. In fact, the theory of natural selection offers very little in terms of a detailed explanation for mankind's existential situation as an animal with self-awareness. From a materialist perspective the "evolution" of consciousness still remains a baffling mystery, as does the enigmatic and sudden appearance of language, race, and culture.

Since its minute, incremental steps are impossible to conceptualize in detail, the evolution drama is, by necessity, a panorama. It is, and can only be, an outline of a shadowy metamorphosis from *animal in-the-world* to Overlord of all planetary life forms. The evolution 'story' dramatizes the 'natural' transfiguration of mankind through a linear procession of metamorphoses which eventually separate him from the animals of his ancestry. Evolution is Western man's totem. Various worldwide creation myths illustrate a similar motif; but, as a scientific theory there is very little concern over the missing details. This is where its faith-based attributes are most evident.

In order to illustrate the faith-based dimension of this theory, it is important that the concerns of the National Academy of Sciences are addressed rationally on both sides of the argument. Therefore, the so-called 'supernatural' should include any invisible force that purportedly drives evolution in the direction of greater complexity, consciousness or ultimate end; or, for that matter, any direction at all. Therefore, the same theories which are attempting to force a square peg (Darwin) into a round hole (the fossil record) should be scrutinized for their 'supernatural' underpinnings as well.

In keeping with the proclamations of Earth's Academies and Courts and other Regulatory bodies, the paradigm of natural selection is the only explanatory route which remains after official slicing and dicing of deductive reasoning cuts out the elusive 'super' in natural. But the Empire's empiricism on this count is peculiarly lax. There is no plausible theory which can support an empirical test of the elusive 'natural'in 'selection.'Most scientists, Phillip Johnson asserts in *Darwin on Trial*, are simply looking for any kind of "confirmation of the only theory one is willing to tolerate." For those of you who haven't noticed, this is not science.

Larmarckian evolution, the precursor to Darwinian theory, posited that species change because they have a desire for a certain feature, and that the organism purposefully changes and passes those changes on to its offspring. Genetic research has disproved Lamarck and has shown that the genes cannot be affected by "the will of the individual." Yet, it will be shown that hints of Lamarckian evolution are interlaced throughout popular evolutionary explanatory 'tales.'Darwinian evolution claims there is no inheritance of acquired characteristics, and posits instead a mechanistic and more or less accidental process of natural selection. As we will see, true Darwinian evolution is rarely exemplified in popular writings, and we must learn to decipher Lamarckian from Darwinian assumptions.

The Evolution Feud

In his well known books and articles on evolution, popular science writer Stephen Jay Gould has attempted to steer Darwinian theory away from natural selection as the lone process involved in evolution. A 10/3/97 *Boston Globe* article entitled, "Survival of the theorists," outlined the crux of the argument within the evolution and evolutionary biology academic factions. The article quotes Gould as saying that "too many biologists, psychologists, and philosophers are buying the notion that natural selection is the be-all and end-all of evolution." He warns that this situation is "bad for science" and, further, is "fueling the growth of evolutionary psychology, a field full of 'narrow, and often barren speculation' about how and why humans behave as they do."

"In a sort of modern-day Darwinian adaptation," proclaims *Globe* journalist John Yemma, "sociobiologists evolved into evolutionary psychologists and animal behaviorists in order to survive the intellectual onslaught." Gould asserts that this way of seeing evolution "puts natural selection on a pedestal not even Charles Darwin would have wanted it on." Addressing one of these evolutionary psychologists, Daniel Dennett, Gould described Dennett's faction as "Darwinian fundamentalists" with a "propensity for cultism and ultra-Darwinian fealty." He further assessed Dennett's book, *Darwin's Dangerous*

Idea, as an "influential but misguided ultra-Darwinian manifesto."

In response, Dennett argued that Gould has created "artificial distinctions." He claimed that, because Gould is such a prolific and capable popular science writer, "the public may be getting misled into thinking there is fire beneath all the smoke he is blowing." Dennett asserted that the public needs to know that Gould's views are not widely shared by evolutionary biologists. Could he be taking heat for labeling the "extreme rarity of transitional forms in the fossil record" as "the trade secret of paleontology"?

In a review of Dennett's book, British biologist John Maynard Smith stated that most evolutionary biologists see Gould as "a man whose ideas are so confused as to be hardly worth bothering with." The reason that this faction had not attacked Gould earlier than this, Smith added, was because they figured he was "on their side against the creationists." The author of the *Globe* article, Yemma, asserts that "depending on whose argument is being made here, there *may be crucial scholarly distinctions* at stake. It is hard to tell." If it's so hard to tell, the *Globe* should have put someone else on the story. Puffing himself up like a blowfish, he adds that "the public could be excused for seeing this as one of those perplexing academic arguments that in an earlier age would have involved angels dancing on the head of a pin."

Why should the public be excused from understanding the basis of this simple argument? Why couldn't this author have explained the argument, even in abbreviated form? Is it because he doesn't understand it himself or because *the media wants to maintain a barren distance* between the public and scientific theory? In effect, what we see on brazen display here is the media attitude that the public is not expected to understand evolutionary theory and is enjoined, instead, to reel around on the head of a pin until confusion sets in and they have to sit down.

Finally, Yemma writes, "just in case creationists are listening in, all parties take pains to point out that this fight has nothing to do with God, religion, the Bible or, as Gould put it, *attempts to smuggle purpose back into biology.*" It is, the contenders say, "an argument well within the world of secular science." Apparently this writer thinks that "creationists" can't read the newspaper, and those who can, he bargains, will be unable to see through his smug coverage of this important topic.

How could this argument possibly not have anything to do with God or religion? There is no getting around the fact that the evolution tiff is a war between atheist and religious contingents. Atheism is the zeal behind all of this spunky rhetoric. I can personally attest to the fact that atheists actually get *high* on Darwinian dogma. It is nothing short of Acada-Media mind control. The mind-numbing fear of all the principals involved in this 'survival of the theories' is based on the fact that the evolutionary record is, as we shall see, incompatible with Darwinian natural selection and compatible with purposeful design. Clearly, it is just this "smuggling of purpose" into evolutionary theory that is *the devil* to the Hatfields and McCoys of Evolutionary Theory for, as we shall see, it is the only truce for which they are willing to put down their shot guns.

With regard to this ongoing feud in evolutionary science, Stephen Jay Gould wrote in *The New York Review* that "we will not win this most important of all battles if we descend to the same tactics of backbiting and anathematization that characterize our true opponents." The "true opponents" of this atheistic bunch are obviously religious creationists, but let's widen the fray as we draw that line in the sand to include all BIPEDs (Beings for Intelligent Purpose in Evolutionary Design), those who have the feeling that 'we didn't get here from there' and are experiencing a little Darwinian Dissent. To arm ourselves for this gentleman's duel, let's zoom in on the head of that pin.

The Shape of a Seductive Idea

In his book *Darwin's Dangerous Idea*, philosopher Daniel Dennett tries to downplay typical feuds such as the one portrayed in the *Globe* article. He contends that the "relatively narrow conflicts" which have arisen among theorists have been blown out of proportion and seriously distorted. With regard to Gould's statement in the *Globe* to the effect that evolution adherents need not lower themselves to the level of feuding to which the creationists have crawled, Dennett's attitude toward non-believers is telling. He states: "anyone today who doubts that the variety of life on this planet was produced by a process of evolution is simply ignorant—inexcusably ignorant, in a world where three out of four people have learned to read and write."

So, if you know how to read and write, you *should* know that the prevailing worldview is Darwinian evolution and you would be stupid, rather, inexcusably ignorant, to argue the fine points. Needless to say, Dennett is sure that no controversy could affect Darwinism, which is about as "secure as any idea in science." If science is all about security, it is no wonder the Brookings Institute came to the conclusion it did with regard to the theoretical effect of the discovery of extraterrestrials on the scientific world and on scientists themselves. They concluded it would scare the pants off them.

It would appear that the aim of the Brookings study was not to protect Earth people or religious institutions, but to protect the scientific establishment, i.e. Darwinian evolution. After all, what other discovery could completely shatter the Darwinian mythology of our purely accidental climb out of the ponds of our local habitat Earth?

Dennett states that "Darwin's fundamental idea of natural selection has been articulated, expanded, clarified, quantified, and deepened in many ways, becoming stronger every time it overcame a challenge." In spite of stating emphatically at the beginning of his book that he could provide numerous examples of how the Modern Synthesis has overcome the shortcomings of Darwin's theory, Dennett accomplishes no such feat. Instead, on the last page of *Darwin's Dangerous Idea*, he admits: "I have learned from my own embarrassing experience how easy it is to concoct remarkably persuasive Darwinian explanations that evaporate on closer inspection." Dennett explains that his book has "sacrificed details" in order to provide a better appreciation of the "overall shape of Darwin's idea," proclaiming the truly dangerous aspect as its "seductiveness."

This seductiveness is indeed very dangerous. It is what compels people to fight tooth and nail on the side of an unscientific theory. Dennett insists that natural selection is best explained at the level of a "blind, mechanical and algorithmic process," dependent on chance alone. He explains that the "mindless" steps of Darwin's natural selection are the outcome of "a cascade of algorithmic processes feeding on chance." Thus, explaining natural selection *mathematically* is Dennett's idea of "rising above the microscopic view to other levels, [and] taking on idealizations." Anyone who has "learned to read and write" will know that there is simply no getting around analysis of Darwin's theory at the micro level, unless one is afraid what might be found there. Alluding to "algorithms" is simply an abstraction used to explain another abstraction. Dennett's cascade of abstractions resolves none of the quandaries of Darwinian natural selection.

Dennett states that "the only way to answer questions about such huge and *experimentally inaccessible patterns* is to leap boldly into the void with the risky tactic of *deliberate oversimplification*," asserting that "oversimplified models often actually explain *just what needs explaining*." He also asserts that "when what provokes our curiosity are the large patterns in phenomena, we need an explanation *at the right level*." He adds, "if science is to explain the patterns discernible in all this complexity, it must rise *above the microscopic view to other levels, taking on idealizations* when necessary so we can see the woods for the trees." He deduces *"could anyone imagine how any process other than natural selection* could have produced all these effects?"

The experimentally inaccessible patterns which can only be explained by oversimplified models are part and parcel of the speciation problem. Darwinists have not been able to zoom in on any proofs of the evolution of any one species into another, so instead they construct seductive dramas. Dennett's proof is to maintain that Darwinist theory is so on the mark that it constitutes "a complete reversal of the burden of proof." So, now we need to prove that evolution didn't happen? This preposterous reasoning confirms Phillip Johnson's assertion that most scientists are looking for confirmation of the only theory they are willing to tolerate. "Could anyone imagine" any other explanation for Dennett's peculiar line of logic? To outline the shape of a seductive idea does not describe the practice of science.

The philosophical hoops which dramatize the evolution story may fool most of the people all of the time, but such dramas are actually contrary to currently accepted science concerning natural selection. Evolutionary themes utilize metaphors which describe a vast journey stretching from a distant past to an imaginary future, infused with emotions ranging from euphoria to despair. Such dramas might be laden with emotional reverence for future human beings and their technological prowess, or may fabricate a more fatalistic scenario which carries humanity toward extinction. Why do such dramas attend evolution? Taken literally and without personal meaning, the theory of evolution is hardly within reach of human imagination. While we can express abstractions and terminology which are supposed to describe such a vast cosmological scheme, the 'facts' involved in such a complex theory have very little in common with the present.

Dennett sees Darwin's "dangerous idea"—natural selection—as a universal acid, eating through "just about every traditional concept leaving in its wake a revolutionized worldview." Perhaps this was true in its heyday; but a survey of the valid arguments against Darwin would appear that it's Dennett's worldview that's in trouble.

Darwinian Hindsight

Geneticist Steve Jones has made the remark that "if there is one thing which *Origin of Species* is not about, it is the origin of species." Nonetheless, in spite of the fact that Darwin's manifesto has trouble even defining the concept of species, his followers believe "the fact of speciation itself is incontestable." Of course, winding backward from the fact that species exist, any mechanism whatsoever can be postulated. The practice of Darwinian Hindsight is far from scientific. "Whatever the mechanisms are that operate," writes Dennett, "they manifestly begin with the emergence of variety within a species, and end, after modifications have accumulated, with the birth of a new, descendant species." Beneath this doublespeak lies the simple reiteration that via an unknown mechanism, variety within species eventually leads to speciation. This statement merely repeats Darwin's thesis over a hundred years later. This is progress?

The fact is, Darwin never quite defined his terms. He was unable to securely pin down this process from "well-marked variety" to "subspecies" and on to "well-defined species." As Darwin wrote in *Origin of Species*, "it will be seen that I look at the term species as one *arbitrarily given for the sake of convenience* to a set of individuals closely resembling each other, and that it *does not essentially differ from the term variety*, which is given to less distinct and more fluctuating forms."

Darwin's attitude throughout *Origin of Species* is that "varieties" are simply "incipient species." Forever teetering on the edge of potentiality, species are always in a hapless phase of *becoming*. This suspension of actuality is the Darwinist way of non-explanation. How have we based an entire cosmological scheme on such ill-defined terms? Darwin never purported to explain the origin of the first species, or the origin of biological forms, or of the Universe itself. He merely began in the middle and tried to work his way back utilizing a circular motion inside of a box. These are the foot prints which all Darwinists seem to follow, for this is the only methodology possible.

The enclosure surrounding the natural selection tautology does not seem to bother most Darwinists as they respond to criticism with rhetorical statements aimed at a person's educational level. In this case, the education itself is nothing more than the indoctrination of a pervasive materialist mindset within the confines of a "specialist" caste system. But, tautologies in scientific paradigms are not new to Thomas Kuhn, author of *The Structure of Scientific Revolutions*. Kuhn assures us that such circular arguments typical of scientific paradigms cannot be made logically compelling "for those who refuse to step into the circle." It would appear that this oddity of science is an enigma explainable only by the well-known motto: 'For those who believe, no explanation is necessary, For those who do not believe, no explanation is possible.'

If Darwin himself never quite defined his terms, how can we be sure we are talking about the same thing? We can't. The only fully agreed upon definition of "species" in *Origin of Species* is Darwin's discussion of "reproduction isolation," the inability of groups to interbreed. Problematically, interbreeding would re-unite groups which are ostensibly in the act of splitting apart genetically, thus frustrating the process of speciation, if such an event occurs at all. As Dennett notes, "if the irreversible divorce that marks speciation is to happen, it must be preceded by a sort of trial separation ... ." Dennett admits that "the criterion of reproductive isolation is vague at the edges."

The Fitness Test

The idea of natural selection is fundamentally different from artificial selection or breeding. The fundamental assumption of Darwin's idea of natural selection is that it is a process which maintains the genetic fitness of a population by ensuring that the most fit individuals survive to produce the most offspring. Pay particular attention to the terms *fit* and *offspring*. A biological species is a group which is capable of interbreeding to produce viable offspring; that is, offspring which can reproduce. The breeding of a new or distinct species which is incapable of reproducing does not constitute a viable species. Creatures who do not survive to produce offspring do not supply the gene pool with their genes which, we may *presume*, were somehow deleterious rather than genetically advantageous or fit. But we are simply making presumptions after the fact. Darwin's concept of natural selection simply defines the fittest as the individuals which survive; the fittest organisms are, plain and simple, the ones which produce the most offspring.

Evolutionists argue that Darwin never claimed natural selection to be the exclusive mechanism of evolution. Selection merely preserves or destroys something that already exists. Mutation must provide the innovative changes in design which natural selection then tests out in the field. Luckily for Darwinists, mutations come in all sizes. Mutations which are large enough to cause visible and immediate changes are deleterious to the organism. Darwin once wrote to his contemporary Charles Lyell that "I would give nothing for the theory of natural selection if it requires miraculous additions at any one stage of descent."

Mutations, as Johnson explains, are "randomly occurring genetic changes which are nearly always harmful when they produce effects large enough to be visible," but which may "occasionally slightly improve the organisms ability to survive and reproduce." Yet, Darwin asserted that this force of natural attrition is also responsible for crafting, over billions of years, the variety of life forms on planet Earth. Darwin proposed that, given enough (1) time and (2) sufficient mutations and variations of the right sort, complex organs as well as adaptive behavior patterns could be produced in incremental steps without outside guidance, intelligence, ultimate goal or purpose. Since Darwin did not have any examples of natural selection with which to illustrate his assertion, he used examples of artificial selection or breeding under the presumption that the same process was at work. But Darwin's analogy to artificial selection, Johnson points out in *Darwin on Trial*, is problematical in many aspects. He argues that "plant and animal breeders employ intelligence and specialized knowledge to select breeding stock and to protect their charges from natural dangers. The point of Darwin's theory, however, was to establish that purposeless natural processes can substitute for intelligent design."

Mutation is defined as a set of mechanisms which provide the genetic variation for natural selection to go to work, including those which we won't go into detail here: point mutations, chromosomal doubling, gene duplication, and recombination. What is important here is that, according to Darwinian theory, variations are supposed to be random; no guiding force causes advantageous mutations at the right time toward any particular end result or more complex form.

Breeders, on the other hand, produce variations in genes "for purposes absent in nature." If breeders were interested in 'survival of the fittest' only, such extremes in variation would not exist such as are evident in the dog world, for instance. Therefore, in the real world, natural selection appears to be "a conservative force that prevents the appearance of extremes in variation that human breeders like to encourage." In point of fact, domestic animal breeders have produced no new species; the new offspring are always capable of interbreeding with the parent gene pool. The results of artificial selection are actually powerful testimony against Darwinian evolution. The fact is, writes zoologist Pierre Grasse as quoted in *Darwin on Trial*, "selection gives tangible form to and gathers together all the varieties a genome is capable of producing, but does not constitute an innovative evolutionary process."

Natural selection presupposes that the fittest organisms are the ones which produce the most offspring. We can presume a characteristic to be an advantage because a species which has it (wings, eyes, large brain, claws, fur, bipedalism, language, etc.) seems to be thriving, but it is impossible to identify the particular characteristic or advantage which has produced the coveted outcome of survival. In Darwin's theory, advantage means nothing more than success in reproducing, or increasing the population for survival of the species as a whole. We can surmise, then, that the individuals which survived to produce the most offspring are doing something right, but that is all we can do. We do not know, specifically or empirically, what they are doing right, but we presume that they must have had the qualities required for producing the most offspring. Therefore, such assumptions always rely on a bizarre retrospective stance (i.e. it must have been the fur that made the grade, or it must have been the large brain, etc.). Problematically, there is no way to test these hypotheses.

In addition, hidden within the natural selection hypothesis is a meaningless tautology which essentially states that those organisms which 'leave the most offspring, leave the most offspring.' Darwin's fitness test is an all-inclusive tautology which sits in a box by itself, in its own world, and explains nothing outside of this box. All of its assumptions are, therefore, true since they cannot be tested empirically. Furthermore, it is always a truth that in any population some individuals will leave more offspring than others, whether the population is not changing or is headed for extinction. It is also important to note, species would actually change more if the "least favored individuals most often succeeded in reproducing their kind." Natural selection, therefore, while seeming to be a theory which supports genome variety, may in actuality result in narrowing the possibilities of variation. As a matter of fact, the prevailing character of the fossil record just happens to be stasis.

In his book *A New Science of Life,* Rupert Sheldrake has written that "the evolutionary changes which have actually been observed over the last century or so for the most part concern the development of new varieties or races within established species." The most infamous instance is the notable emergence of a darker variety of European moths in an industrial area in England. The "peppered" moths were favored by a mechanism termed "natural selection" because they were better camouflaged against the sooty background, and the birds couldn't see them as well. Anyone who has studied biology in high school or college is aware of this well-worn Kettlewell's peppered moth experiment, the classic demonstration of natural selection which is obviously not a demonstration of the origin of any species. It is merely an example of variation and fluctuation in a local gene pool due to environmental factors. Due to the fact that the darker moths survived to produce more offspring, more darker moths began to show up in the area since these genes became dominant in the gene pool. The variation in color was simply one of the possibilities of their genome. They did not become another species. In the end, they were still moths, peppered or not.

There is, in fact, no evidence which confirms the hypothesis that the concept of natural selection is an evolutionary process capable of producing innovative designs in organs and organisms. In fact, states zoologist Pierre Grasse, such "proofs" of evolution-in-action are simply "observation of demographic facts, local fluctuations of genotypes and geographical distributions." Such fluctuations, he asserts, do not assert an innovative evolutionary process.

As John Davidson writes in *The Web of Life*:

> Evolutionary theory presents one of the most explicit examples of *a priori* reasoning, and even blind faith, ever seen in a supposedly scientific hypothesis. Books on evolution are full of the prior assumption that evolutionary theory is correct. The facts are then presented to fit the theory. And although many other interpretations of these facts are also possible, it is a rare biologist who dares to be a dissenter or to even suggest that other interpretations and explanations are also possible.

The Whole and Its Parts

Darwin was, in effect, a gradualist, believing that every major transformation in form was the end result of a cumulative process of incremental change and adaptation. As Philip Johnson points out, Darwin asserted that natural selection was a process of "preservation and accumulation of infinitesimally small inherited modifications, each profitable to the preserved being."

Darwin's theory emphatically avoided any leaps or jumps in evolution, called "saltations," which resulted in a new species in one generation. Such a leap being equal to a miracle, or an act of creation, Darwin asserted that he would have to throw out his baby with the bath water were it ever proven that evolution required saltations, or systemic macromutations as they are called today. Systemic macromutations are considered theoretically impossible today, since complex assemblies of parts cannot change simultaneously as a result of random mutation, that is, in a preserved being. Such a large and visible occurrence of mutation would be murderous to the organism.

In the last fifty years, biochemists have begun to decipher the enormous complexity within cellular structures, which incorporate sometimes hundreds of precisely tailored processes. With the increase in the number of required parts of a system, the impossibility of a gradual scenario skyrockets. Complex entities don't evolve piece by piece, asserts Michael Behe, they have to be designed from the start. In his book, *Darwin's Black Box: The Biochemical Challenge to Evolution*, Behe outlines a number of biochemical systems, such as cilium, flagellum and blood clotting, that cannot be explained by Darwinist gradualist explanations.

For instance, Behe writes, if the shape of a protein is warped, it simply fails to do its job. Specifically, the shape of a folded protein and the precise positioning of the different kinds of amino acid groups allow the protein to work. If the job of one protein is to bind to another specific protein, Behe explains, their two shapes must fit each other precisely. If there is a positively charged amino acid on one protein, it will fit only with a negatively charged amino acid. If it is the job of a protein to catalyze a chemical reaction, the shape of the enzyme must match the shape of the chemical target. In addition, Behe explains, enzymes have amino acids precisely positioned to cause chemical reactions.

In short, the work of every cell in the body requires teams of proteins, made up of amino acids, and each member of the team carries out just one part of the task. Not one of these chemical reactions is allowed to go awry in a functioning system. Behe concludes that irreducibly complex systems cannot evolve in Darwinian fashion. The whole system has to be put together at once. He states: "You can't start with a signal sequence and have a protein go a little way towards the lysosome, add a signal receptor protein, go a little further, and so forth. It's all or nothing." In his analysis of complex parts of various biological systems, Behe concludes that "it is extremely implausible that components used for other purposes fortuitously adapted to new roles in a complex system."

Human and animal bodies contain an array of interrelated systems containing organs, tissues and chemical components in intricate order. How would it be possible to build into this system random micro-variations during each tiny step which are at the same time *profitable to the preserved being*? Surely some of the these incremental changes would be detrimental at some place along the way to the cumulative result, which is at the same time supposed to have no goal toward greater complexity. Furthermore, such infinitesimal changes would not necessarily be of any immediate advantage unless other parts needed for it to function also appeared with it. What we need to imagine here, Phillip Johnson points out, is "a chance mutation that provides a complex capacity all at once, at a level of utility sufficient to give the creature an advantage in producing offspring."

Problematically, since macromutations are always maladaptive, Darwinists assert that complex and similar organs must have evolved independently, over and over again in many different organisms, by the accumulation of tiny micromutations over a long span of time. One example is the evolution of the eye. Did the eye evolve separately at first, and if so was it useful for some other purpose other than vision? Did the neural capacity for vision evolve in incremental steps along with the eye? What good is 5% of an eye, and what good is any percent of it without the neural capacity to process the information it records? Evolutionary biologists use the fossil record to indicate a plausible series of intermediate eye designs, but the problem is the designs belong to different animals and involve vastly different types of structures (some having just a pinhole eye with no lens or some being set in a cup, for instance) rather than a similar structure which added to its complexity over time. There is no evidence that it is structurally the same eye design at all.

Furthermore, it has been noted that no fossils of animals which now exist have been shown to have something that would indicate an earlier or less complex eye structure. For instance, the nautilus sea creature, given hundreds of millions of years, has not evolved a lens for its eye despite having a retina "practically crying out for this particular simple change." Dawkins is quoted as saying that "virtually all the mutations studied in genetics laboratories—which are pretty macro because otherwise geneticists wouldn't notice them—are deleterious to the animals possessing them." In order to pass all of these tests simultaneously, followers of Darwin have "evolved an array of subsidiary concepts capable of furnishing a plausible explanation for just about any conceivable eventuality," states Johnson.

Life is more than chance combinations of atoms and cells, write the authors of *Giants of Gaia*. To organize the parts which "collectively enable a bird to fly, or the human brain to form, the writers insist, there had to be an order which brought together the parts not by chance, nor by simple adaptation to external stimulus, but through intelligence." This intelligence inherently constitutes the Universe, or Mind-at-Large.

Punctuated Equilibrium

It has been noted by paleontologist Niles Eldredge that certain restrictions make it difficult to pursue a successful "career" as a Darwinist. Ironically, those restrictions arise from the fossil record. He writes that the pressure for positive results is enormous. The various schema which these stressed-out researchers must juggle is Darwin's insistence on gradualism on one hand and, on the other, the findings in the fossil record which point to saltation, as well as catastrophism. Johnson quotes Eldredge in *Darwin on Trial*: "either you stick to conventional theory despite the rather poor fit of the fossils, or you focus on the empirics and say that saltation looks like a reasonable model of the evolutionary process—in which case you must embrace a set of rather dubious biological propositions." (Johnson, 60)

Thus, it is clear that paleontologists who are tethered to neo-Darwinism are not free to draw apt conclusions to which their "dubious" evidence points. In order to operate within the neo-Darwinist boundaries and at the same time achieve success with their projects (and, therefore, future funding and paychecks) another subsidiary theory called "punctuated equilibrium" was hatched. This theory posits that organisms remain the same over long periods of time and that evolutionary changes take place rather abruptly. Punctuated equilibrium is actually an attempt to strike a balance between what Darwin hoped would be discovered in the fossil record and what has actually been found since 1859. How different is punctuated equilibrium from saltation or creation? The embarrassing fact is that, despite an enormous amount of interim fossil hunting, according to Stephen Jay Gould: "the history of most fossil species includes two features particularly inconsistent with gradualism." Those two features are stasis and sudden appearance.

Gould wrote that most species exhibit no directional change during their tenure on Earth and that they appear in the fossil record *looking morphologi - cally the same as when they depart*. He also wrote that species do not arise in a local area by steady and gradual transformation but, rather, *species appear all at once and fully formed*. Yet, in spite of the fossil record essentially displaying saltation, Gould and other neo-Darwinists remain devout apologists for the theory of natural selection. Johnson succinctly states the problem in *Darwin on Trial*: "natural selection is a guiding force so effective it could accomplish prodigies of biological craftsmanship that people in previous times had thought to require the guiding hand of a creator."

The Systems View

Rather than viewing life as a Malthusian, dog-eat-dog fight for survival, systems thinkers view it as a cooperative enterprise. Systems theory asserts "continual cooperation and mutual dependence among all life forms as central aspects of evolution." Systems thinkers contend that life took over the Earth through "networking" rather than by fighting tooth and claw for independent niches.

Microbiologist Lynn Margulis asserts that neo-Darwinism is based on outrageously outdated reductionist concepts. The common picture of the genome as a linear array of independent genes corresponding to a biological trait is being called into question. A single gene may actually affect a wide range of traits, and separate genes may combine to produce a single trait. Systems theory sees the genome as a biological network. Stating that complex structures could not have evolved through successive mutations of individual genes, Margulis concentrates on the 'coordinating and integrating activities of the entire genome.'

Margulis has also argued that practicing neo-Darwinists, most of whom come out of the zoological tradition and deal with a relatively recent part of evolutionary history, lack relevant knowledge in microbiology, cell biology, biochemistry and microbial ecology. Importantly, she states that "current research in microbiology indicates strongly that the major avenues for evolution's creativity were developed long before animals appeared on the scene."

In *Web of Life*, Fritjof Capra quotes Margulis: "when scientists tell us that life adapts to an essentially passive environment of chemistry, physics, and rocks, they perpetuate a severely distorted view. Life actually makes and forms and changes the environment to which it adapts. Then that 'environment' feeds back on the life that is changing and acting and growing in it." James Lovelock, the author of the Gaia hypothesis, writes: "So closely coupled is the evolution of living organisms with the evolution of their environment that together they constitute a single evolutionary process."

Vitalism and the Gaia Hypothesis

It was Margulis, along with Lovelock, who formulated the Gaia hypothesis in the 1970s. They proposed that life creates the conditions for its own existence, thus challenging the reigning theory that the forces of geology set the conditions for life and plants and animals, sort of accidentally along for the ride, evolved by chance under the right conditions. The Darwinian concept of adaptation to the environment has been seriously questioned by Margulis, Lovelock and others working from a systems point of view. Evolution cannot be explained by the adaptation of organisms to local environments, they argued, because the environment is also being shaped by an overarching and cyclical network of living systems. The evolution of life according to the Gaia hypothesis is a cyclical, "self-regulating" feedback relationship.

Proponents of 17th century vitalism posited that the body is governed by the action of a soul or vital force. This teaching asserts that evolution is not purely mechanical but is the result of a purposive force called the 'life force' which pervades the Universe. For instance, the vitalist T.E. Hulme wrote that "the process of evolution can only be described as the gradual insertion of more and more freedom into matter... In the amoeba, you might say that the impulse has manufactured a small leak through which free activity could be inserted into the world, and the process of evolution has been the gradual enlargement of this leak." (*Beyond the Outsider*, 121) Neo-Darwinists argue against such an ulterior impulse, vital force or purpose.

The vitalist school of thought argues that physics and chemistry are insufficient to explain life. The whole is more than the sum of its parts: this is what vitalism has in common with systems theory. In addition, both vitalism and organismic biology are opposed to the reduction of biology to physics and chemistry. Both vitalists and organismic biologists try to describe the way in which the whole is more than the sum of its parts. While organismic biologists view the inherent relationships which organize the whole as being the presumed added ingredient; vitalists look for an outside force, field, or nonphysical process. A modern example is Rupert Sheldrake's theory of morphogenetic fields, described in detail in his book *A New Science of Life*.

In systems thinking, properties of an organism or living system are properties of the whole which none of the parts alone exhibit. According to systems theory, the properties of the whole arise from the relationships among and interactions between the parts, for such properties do not exist when the parts are isolated. The nature of the whole functioning system is qualitatively different from the mere sum of its parts. This is directly contrary to the reductionist approach, in which parts are analyzed by further and further dissection, analysis and reduction.

Contemporary organismic biologists describe a "system" as a highly organized network of feedback loops arranged in varying levels of complexity. They see no need for a separate, nonphysical concept operating outside of the patterned relationships of physical structures, because they are of the mind that these patterns are "self-organizing." Thus, where vitalists see an outside force or entity as designer or director, modern systems thinkers see merely a pattern of self-regulation arising in nature. How far does the concept of "self-organizing" go to actually explain the quality of the whole being more than the sum of its parts? To say that something is "built-in" does not prove there was no builder. We still have Arthur Koestler's "ghost in the machine." What constitutes the "self"?

To deduce that an internal design is "self" regulating does not fit into the Darwinian paradigm. Therefore, we are back at the beginning. How is the whole more than the sum of its parts? It is just this idea of "self-regulation" which gave the Gaia theory the boot by the scientific community when it was initially proposed. They rightly queried how life could create and regulate the conditions for its own existence without bringing into play a purposeful overriding force. The idea of natural processes being in any way guided was unscientific because it was *teleological*. Yet, keeping carefully within certain necessary aspects of evolutionary theory, Margulis and other systems thinkers continue to assert that there is no purpose or over-arching goal in evolution, stating that the *driving force* of evolution is not random, but rather emerges out of "life's *inherent* tendency to *create novelty*, in the *spontaneous* emergence of increasing complexity and order."

'Tis a good thing that this *creative force* is *inherent* and *spontaneous*, otherwise it wouldn't square with the naturalist paradigm addressed by the Supreme Court. Clearly, it's one thing for the new wave of systems thinkers to partially debunk Darwin, but they had better stop short of saying the driver is anywhere but *inside* the vehicle! You can toss out the bath water, but this Baby is 'on Board'! Keeping carefully within certain necessary aspects of Darwinian theory, Margulis and other systems thinkers explicitly assert that there is no purpose or over-arching goal in evolution.

If evolution is the gradual change of one kind of organism into another kind, then the fossil record, point blank, indicates that evolution has not occurred. It is not difficult to see that evolution has achieved the status of a religion in western society. Evolution is a powerful creation myth that shapes our view of who we are, and influences us in ways far beyond its official function as a biological theory.

In her book *Evolution as a Religion*, Mary Midgely asserts that "the theory of evolution is not just an inert piece of theoretical science," but is also "a powerful folk tale about human origins." She warns against applying the confidence due to well-established scientific findings to a "vast area which has only an imaginative affinity with them," and where only the "trappings of a detached and highly venerated science are present."

Sociobiological Motifs

It is most important to learn to recognize sociobiological motifs hidden within evolutionary philosophies. For instance, Sir Julian Huxley has written: "As a result of a thousand million years of evolution, the Universe is *becoming conscious* of itself, able to understand something of its past history and its possible future. This cosmic self-awareness is being realized in one tiny fragment of the Universe—in a *few of us human beings...* The first thing that the human species has to do to prepare itself for the *cosmic office* to which it finds itself appointed is to explore human nature, to find out what are the possibilities open to it (including of course its limitations)." In *Beyond the Outsider*, Colin Wilson adds: "Man has a choice; he can devote himself to evolutionary purposes, or confine himself to his everyday animal purposes."

Evolutionary philosophy is replete with sociobiological allusions to such things as "evolutionary purposes" and "cosmic offices." Yet, Darwin was specific in his denunciation of any such overarching tendency. It would appear that the human species has simply appointed itself to this office.

In addition, Huxley writes that the *attainment of greater complexity* in the forms of life denies the law of entropy, in that, while "the Universe of physics is running down; the Universe of evolution is winding up." Huxley asserts: "...on this planet the second law of thermodynamics is now not working, and of course [this] opens up the possibility that *there may be agencies operating in the Universe supplying energy* which would enable the whole cosmos to behave in an anti-entropic manner."

To this we should argue that there is no such "greater complexity" in forms of life. All forms are *equally complex* in their own right. Additionally, modern mankind may be technologically evolving, but *physically*, as we will soon discuss, Jack Cuozzo's work with Neanderthal skulls suggests mankind may actually be *devolving*. In addition, who says the Universe has become aware of itself only after "a thousand million years of evolution"? How do you know it wasn't aware of itself all along? Is it possible that we see various systems as "evolved" simply because we're stuck in a linear concept of time? Additionally, is it not rather strange to view the "evolution" of cosmic consciousness as "being realized in one tiny fragment of the Universe—in *a few of us* human beings..."? Which few might those be? This peculiar sociobiological point of view is isolationism and anthropocentrism at its finest. But it doesn't stop here. Following is perfect example.

With the publication of various popular science books attempting to simplify the new physics paradigms for us "little people," this indulgence in evolution as a creation drama is most obvious. A viewing of the science section of any large book store will display count-

Drawing by Thomas Huxley (1825-1895). Huxley, known as "Darwin's Bulldog," was the first propagandist and agitator for the new science of biology, and wasn't concerned about the mechanism by which evolution occurred.

less titles which seemingly portray the idea that science is making room for the existence of God. It is doing no such thing. It is calling *itself* God.

In his book *The Mind of God: The Scientific Basis for a Rational World*, Paul Davies makes an attempt to redefine God-hood as a process of rational thought which is pervasive in the Universe, having a mathematically recognizable pattern ultimately reflective of human-hood. There is no indication anywhere in the pages of this book that the author is talking about God as the omniscient, omnipotent, and determinant cause or creator of the "rational" Universe inside and outside the human mind. It is, rather, a book about human rational superstructures in the act of recognizing that the way it thinks might reflect the way the Universe was built.

This modern conversion of God goes one step beyond merely creating God in man's image, to creating God in the image of Scientific Prowess—the Buddha of Rationality. Davies writes:

> Human beings have all sorts of beliefs. The way in which they arrive at them varies from reasoned argument to blind faith. Some beliefs are based on personal experience, others on education, and others on indoctrination. Many beliefs are no doubt innate: we are born with them as a result of evolutionary factors.

Buried in this obtuse Lamarckian epistemology lies the suggestion that a certain belief system, an acquired characteristic by any standard, can be considered an "evolutionary factor." Wouldn't it be handy if we discovered that Davies was setting the stage to present the thesis that scientific rationalism is the *correct* belief system of all well-evolved individuals? He is! Davies writes:

> Four hundred years ago science came into conflict with religion because it seemed to threaten Mankind's cozy place in the Universe ... The revolution begun by Copernicus and *finished by Darwin* had the effect of marginalizing, even trivializing, human beings.

Davies wonders why "science works," and asserts that it works so well that it points to something profoundly significant about the organization of the Cosmos. The concept of human reasoning, he explains, is itself a curious one. He writes: "The processes of human thought are *not God-given*. They have their *origin in the structure of the human brain*, and the *tasks it has evolved* to perform. The operation of the brain, in turn, depends on the laws of physics and the nature of the physical world we inhabit ..."

This peculiar evolutionary psychology (i.e. sociobiological) model sets its definition of God as the mechanistic processes in nature which seemingly mirror the belief system of scientific rationalism. As Charles Fort asserts, "science is a Turtle that says that its own shell encloses all things." The author's definition of the conscious awareness of the relatedness or connectedness of inner/outer worlds is the now pseudo-scientific term: "God." The processes which mirror scientific rationalism are now called God; not the giver, mind you, but the gift itself. Yes, ladies and gentlemen, Science *is* God.

In his book, *The Self-Aware Universe,* Goswami has a similar point of view. He asks: "How has the Cosmos existed for the past fifteen billion years if for the bulk of this time there were no conscious observers to do any collapsing of wave functions?" This logic is preposterous. We assume a closed system, based upon the Darwinian paradigm of the local evolution of consciousness on Earth as an independent and accidental event. Secondly, we suppose that consciousness is specifically "human." The most simple answer is that we are not the first or the only conscious observers, and we are not alone in the Universe!

Pondering how consciousness "arose in the Universe," this peculiar Western viewpoint—the anthropic principle—refuses the primacy of consciousness, and instead assumes causality—an endless chain of causes—in the Universe. The anthropic principle assumes the evolution of intelligence from non-organic matter, and extrapolates the time required for the evolution of "conscious observers" based on the presumed localized, one-of-a-kind, anomaly of Earth-based human evolution. This assumption is then applied as a cosmic constant.

The theory of entropy states that the Universe is running down, and as time passes there will be less energy, less organization, and more disorder. A living organism, however, runs counter to an entropic system by consciously forming an internal order. Despite the theory of entropy, the Universe does not appear to be running down. It is more like a living organism, with a conscious purpose. Modern chaos theory sees evidence that the *primal* presence of consciousness is in itself the reason why the Universe is not running down. In other words, consciousness or "Mind-at-Large" is primal to "human" consciousness, not the other way around. We've got it backwards. It's tricky though, when, according to *Giants of Gaia*, new physics paradigms tend to show that the Universe has "an obliging nature" and can "provide proof for any cosmological scheme, scientific or mystical, foisted upon it."

What Davies has taken so many pages to say is that mankind has evolved in a Universe that looks like himself; that walks like him and talks like him. Having grown inside of the womb of the Universe, mankind is "star stuff," to use Carl Sagan's term. These familiarly quaint ideas do not prove that the process of human rational thought is *not God-given*. Davies indicates that humankind has acquired a certain "correct" point of view with which to peer into a micro and macro Universe of unfathomable clockwork, while at the same time arguing against an ultra purpose or design. Evolution becomes a grand cosmological scheme in which man "evolved" patterns of thought which correctly mirror the Universe. Mankind is a being which "accidentally" hit the bulls eye.

Since evolution is the only theory they are willing to tolerate, Darwinists have allowed an untestable hypothesis to "finish the revolution," thus cutting off all true and open inquiry into the nature of the Universe and whether it is a manifestation of a prime mover or a vital cause. In William James' terms, we have outlawed alternative descriptions of our world which we find inconvenient to our preconceived ideology. Colored by the attitude that scientific materialism is the be-all and end-all, that the revolution is finished, that all of life is reducible to a physical component that was 'there'in the beginning, we shall look straight into the face of our Creator and never see the view. But what was it that was there in the beginning?

The Escalator Myth

A consistent pattern in popular evolutionary theory is that evolution progresses 'upward' toward more complex forms. This is actually contrary to Darwinian theory. For instance, Peter Bowler, author of *The Non-Darwinian Revolution: Reinterpreting a Historical Myth*, finds no fault with Darwin's theory; he only finds fault with "the mistaken notion of its revolutionary effect on nineteenth-century thought." Examining the work of such figures as Owen, Spencer, Kelvin, Huxley, Haeckel, and Freud, Bowler finds "a near-universal tendency to accept evolutionism while rejecting Darwin's central premise: natural selection." Therefore, it isn't Darwin at all who has affected modern philosophy, since most philosophers have misapplied the essence of Darwin's theory.

The idea of the upward movement of life forms from lifeless matter through plants, animals and man was suggested by Lamarck and initially given the term 'evolution' by Herbert Spencer. Darwin argued against the idea that there existed any innate tendency toward progressive development. Darwinian theory, instead, was shaped rather like a bush than a ladder, and accounted for all types of development, including unchangingness and regression, as responses to environment. In actuality, Darwin posited no guarantee of the continuation of particular changes, and saw no particular change, such as increased intelligence, which stood at the apex of this metamorphosis. Thus, Mary Midgely deduces in *Evolution as a Religion*, Herbert Spencer's ladder theory has prevailed over Darwin's bush theory in the public mind and in popular and scientific writing, as well as in the minds of scientists who have had a difficult time fitting Darwinian theory to the actual fossil record.

It is also important to note that the human-centeredness of the escalator model not only distorts evolutionary theory, but shapes our attitude toward nature and the other biological life forms with whom we share our world. What gives us the right, Midgely asks, to consider ourselves the "directional pointer and aim-bearer" of the evolutionary process? The originator of the escalator concept, Larmarck, considered animals to be "behind man, engaged on the same journey;" a still persistent notion in Neo-Darwinist philosophy which serves to maintain the inferiority and expendability of animals. This idea supports the attitude that anything outside of our own experience of life has no value.

Such prophetic evolutionary tales exalt certain ideals by projecting them on the screen of a distant future—"a fantasy realm devoted to the staging of visionary dramas." Such dramas, Midgely contends, are based on the moral convictions of the author of such stories, and of the age in which they are born, rather than on truly scientific theories. Midgely suggests that an "over-ambitious reliance on the escalator model and the inflated creeds which express it" are the source of many superstitions which follow the theory of evolution.

Evolution's Panchestron

The philosopher William James defined the religious as an attitude which is "directed to the world as a whole, and about which there is something solemn, serious and tender." A religious attitude is an attitude of acceptance founded upon belief in an unseen order, and which surrenders to a 'larger power' in the sense that 'all things work together for good.' In his book, *Ishtar Rising*, Robert Anton Wilson describes a "panchreston" as a system that explains everything. He argues that "any human formula which explains all human formulas is technically in the class of all classes which include themselves and leads to logical contradictions." How different from such a religious attitude is the "panchestron" of evolutionary theory's Universe and man's place within it?

As Midgely asserts, the myths and dramas attending the theme of evolution, while using scientific language, are "quite contrary to currently accepted scientific doctrines about it." They provide their adherents with a "live faith" which adds meaning to their lives. In this sense they are religious. They are highly charged with a tone that ranges "from the euphoric to the despairing." In these dramas, triumph might be tinged with an air of reverence for the future human beings, or they might contain elements of "brash technological conceit." In the more fatalistic scenarios, a malignancy such as a "selfish gene" or the inability to care for the world will bring humanity to the brink of extinction.

Why do such dramas go hand in hand with evolution? During my years as a believer in the "fact" of evolution, the two foremost responses I tended to give to people who contested their great ape lineage were (1) that they did not have an adequate understanding of the theory (i.e. they were dumb), and (2) that humans were generally incapable of imagining the incredible span of time which would be involved in such incremental processes of change (i.e. it *must* be happening even though we can't see it or prove it). Therefore, it is understandable why Midgely would suggest that, taken literally and without personal meaning, the theory of evolution is "scarcely graspable at all by the human imagination."

While we can express abstractions and terminology which are supposed to describe such a vast cosmological scheme, they actually mean very little to us unless we are the players in the drama. The "facts" involved in such a complex theory have very little in common with the present. Nonetheless, the creators of such evolutionary dramas owe to their readers a more honest exposé of our current understanding of evolutionary processes so that they understand fully what they are accepting as scientific fact. They would quickly realize that the Emperor wears no clothes.

The Role of Chance

Examples of such scientific writers pontificating about evolutionary ideals abound on the library shelves. It is important to realize that, while these authors might have some kind of credentials (i.e. letters following their names), the reader is always at risk of being hoodwinked by literary drama in the guise of scientific fact. An exceptionally interesting example is *The Origins of Life* written by psychiatrist, H.V. Ditfurth. Ditfurth asserts that the *role of chance* which Darwin assigned to evolution has been overstated by the opposition in order to 'trip it up'. Ditfurth suggests that "if chance were the sole driving force behind biological development, not one single functioning organism would have ever come into existence ... everything would have been doomed to total chaos." Rather, he explains, Darwin's theory describes the collaboration of random elements with natural laws, with mutation acting as the *chance* element and natural selection as the *law*.

Ditfurth claims, with apparently straight face, that an increase in the rate of mutation would "undoubtedly hasten the course of evolution, but at a certain point it would begin to jeopardize all further development because too much experimenting would be going on in each generation." Thus Ditfurth is suggesting that nature is "experimenting" rather than producing random copying errors. It is important to note that such language describing nature as a purposeful shaker and mover is commonplace in evolutionary tales and is misleading in the context of the proper understanding of Darwin's thesis. Ditfurth then asserts that a relatively stable species whose rate of mutation suddenly speeded up "would bring forth within only a few generations such an abundance of *crazy variants, monsters, and misbegotten creatures* that it would soon die out from an excessive loss of genetic tradition." He adds, "this seems to have been the fate of some of the *species that disappeared from the face of the Earth* in the dim primeval past." (Ditfurth, 76)

Ditfurth's argument must be taken with a grain of salt, because the fossil record clearly does not show a parade of misfits and genetic monsters representing a chain of mutations that didn't pass the nature test, or whose extinction was caused by "an excessive loss of genetic tradition." In fact, there is general consensus that mass extinctions on Earth have catastrophic causes. If such a chain of missing links actually existed, Ditfurth and others wouldn't need to write evolutionary fiction; the fossil record would speak for itself. But, as we have already shown, what the fossil record is shouting loud and clear is quite different from Ditfurth's "crazy variants and misbegotten creatures." The real data represents saltation, nothing short of the seeming *creation* of a successful parade of viable forms which have "appeared all at once and fully formed."

Since the behavior of elementary particles is unpredictable, mutations are random events, Ditfurth asserts. There is no way of determining which link in the molecular chain will be exchanged for a different one. Mutations are also random in another sense, he points out; that is, they take place "with no regard for the situation of the population whose gene pool they alter." Ditfurth writes that the rate or direction of mutation and a *biological need* for mutation operate in different realms. Whereas the factors determining mutation are on the molecular, micro-cosmic level, so to speak, the environment is the macro-cosmic. The factors on the micro level are completely independent and unaware of changes on the macro level, and vice versa. A mutation, he clarifies, is "blind to the biological situation it helps to decide."

Yet, he explains: "Here too nature has found a way to *wrest a meaning* (*retroactively*) from the whole process." Since information from the macro-world never penetrates the micro-world, he explains, the genome can never learn from experience nor profit from mistakes. Such an assumption would comprise a Lamarckian concept. Yet, he posits, the "resultant blindness of every individual mutation to the situation of the organism with whose blueprint it *mindlessly tinkers* has nevertheless an advantage: this unavoidable blindness leaves the species open to *unforeseeable future possibilities*." (Ditfurth, 79)

This explanation is internally contradictory. At one point mutations are random events with no regard for the situation of the population, and at another point there is a "biological need" for mutation operating on a "different" level. If evolution has no foresight, then it has no "biological need for mutation." If anyone or anything has "wrested" a retroactive meaning from the evolutionary process it is "evolutionary man" not "nature." The peculiar dramas derived from evolutionary theory are always retroactive. Evolutionary theory is not present-centered and cannot be understood in present terms, but is suspended in a past and future mythos. Part and parcel of this mythos is a persistent screenplay which attempts to adhere to the random nature of Darwinian evolution, while at the same time exalting its final product, mankind, retroactively speaking, as the accidental aim and purpose all along. This is Lamarckian epistemology in disguise. And not even a clever disguise, since it reeks of vitalism.

Ditfurth asserts that when environmental conditions change *unpredictably*, "the *wide ranging fantasy* of the mutational principle, which pays no heed to the realities before it, can suddenly prove to be a lifesaver for the species." He explains that "the *mechanism for creating mutations* which was totally absorbed by the concrete demands of the present would be wiped out." Under these conditions, he posits "a *random shot can suddenly become a bulls-eye* no marksman would have managed to hit because there *seemed to be no reason* to aim in the direction that *later turned out to be the right one*."

In the vast majority of cases, Ditfurth writes, the process of evolutionary adaptation would never have caught up with environmental changes "if it simply reacted to them instead of accidentally anticipating them." Ditfurth seems to have forgotten that his stance was supposed to be downplaying the feature of "chance" which, he laments, opponents of Darwin have unfairly exaggerated. Instead, he has done nothing but exemplify the freewheeling glory of "chance" with lingo emphasizing the height of his fascination with the random nature ideal. In addition, Ditfurth seems to forget that the "wide ranging fantasy of the mutational principle" cannot be too wide or it will spell disaster for the species.

The value of a mutation can only be judged after the fact, he writes, for its evaluation depends upon the fitness it conveys to the organism, or as Ditfurth phrases it, "whether it increases or lessens the individual's chances to be one of the select few members of the next generation of parents." This, in turn, depends upon its trial run in the field, or the "effects that mutation has on the organism's performance in the environmental test."

A variable, such as a mutation, which can only be judged after the fact, has no value. This is an absurdity. There are no other variables to manipulate. There is nothing to test. All of the drama attending this explanation cannot hide the fact that this view depends upon only one thing: the pre-formulated and unscientific framework of the human being who is guiding this process of "evaluation." For it is this human being who hypothesizes a certain causal relationship, applies a certain scheme, chooses a certain variable (in this case, warm blood) represented by a certain biological subset which has successfully reproduced.

It is clear that the perceiver is working with knowledge of the end result of the test, and has formulated and proved, in one fell swoop, the hypothesis that this particular variable is the causal agent. It is simply presumed that nature, the ultimate 'scientist', has already performed the field test (which has been given the rather vitalist and contradictory term "accidental anticipation") since the ultimate stamp of approval is simply the survival, or fitness, of the species. The species or subset has survived because it is fit, and is fit because it survived; that is the Darwinian fitness test in a nut shell. It is not science, it is a perverse kind of nature worship in denial of itself. If you doubt it, keep reading.

Thermal Emancipation

Today, Ditfurth writes, the Earth is ruled by "the heirs of a mutation that was originally senseless and that remained a failure for almost half a billion years." That mutation was a metabolic adjustment for a warmer body temperature, which would have been most disadvantageous in the waters which were constantly warmed by the Sun. Such a genetic alteration for warm-bloodedness, suggests Ditfurth, would have cost the organism extra fuel for a "worthless function," and those individuals would have been eliminated from the gene pool because of their increased need for fuel. But this pointless mutation, he explains, "proved to be highly advantageous at a much later stage of the Earth's history," that is, when "life gradually began to encroach upon dry land." There on dry land, Ditfurth proposes, the mutation was suddenly an advantage to the hapless 'misfit.' He writes: "the higher consumption of nourishment was more than compensated by what the organism gained in independence from temperature variations of the new milieu." Cold-blooded creatures who had previously ruled the Earth found themselves up against a new competitor. Ditfurth's peculiar retroactive mythology concludes that, "the most favored heirs of this thermal emancipation are we ourselves, the human race." (Ditfurth, 83)

In this absurd scenario, the "thermally emancipated" individual already knows who won the race and has chosen the winning sneakers after the fact. Ditfurth's deductive reasoning, couched in Darwinian rhetoric, simply states that "all humans are warm blooded; warm blooded individuals are good survivors; therefore, all humans are good survivors." That's a sound enough deductive argument. But Ditfurth extrapolates much more than is warranted based upon this simple logical deduction; he extrapolates a cause and effect relationship where there is none! In terms of explaining an implied causal relationship between warm bloodedness and the evolution of the human being, Ditfurth has not accomplished his task, since "survival" alone—the fitness test—cannot prove or describe any process at work in the evolution of a cold blooded organism to a warm blooded organism.

Clearly, the fitness premise cannot explain a causal relationship between two variables; but rather extrapolates a connection which is unprovable. There is no way to prove or disprove this fantastic story; only one of countless phantasms written by popular science writers, laden with false analogies, value judgments and doublespeak. This melodrama, and others like it by writers like Gould, Dawkins and Dennett, mean absolutely nothing unless you're having an 'easy hair day' and you need to feel akin to a warrior in an ancient, long-forgotten battle, dramatized by typical phrases like *ruled the Earth, a new com -petitor, most favored heirs,* and *eliminated from the gene pool.*

In her book *Evolution as a Religion,* Mary Midgely calls the concept of survival of the fittest is one of the most "dangerous melodramas" of the modern age. Midgely asserts that this danger arises when writers "less careful than Darwin allow the drama to usurp the factual evidence." The theory, she claims, has been heavily distorted by biases arising from oversimplification of the theory and involves people who are "obsessed by a picture so colourful and striking that it numbs thought about the evidence required to support it." The feelings of superiority and strength conjured by these linguistic feats, in fact, may be the reason that popular science/evolution titles are stacked on bookstore shelves and readily peddled by mainstream publishers. But, what gives humans the idea that they are the culmination point in the process of evolution? Midgely suggests that perhaps "any intelligent species, able to meditate on such things, must in some way think of itself as central in the whole world, because in its own world it is so."

Ditfurth's fantastic scenario is completely absurd; yet it is just one out of countless ridiculous dramas extrapolated by evolutionary theorists from Darwin's vague theoretical nonsense. The real horror is that, while research scientists are held accountable for every scratch they publish in academic journals, evolutionary "philosophers" peddle their muddled hogwash with no accountability.

Atheism vs. Religionism?

Why does the philosophy of evolution have to be a feud between atheists and religionists? Why can't the validity of Darwin's supposedly scientific hypothesis be questioned on its own merits without the assumption that the questioner is coming from a religious point of view? This is the fundamental problem undermining the ability to have a serious dialogue about evolution, as the following story illustrates.

In Fall of 1998, I published a version of this chapter as an article in *Paranoia: The Conspiracy Reader.* I received a letter from an acquaintance in California who said that he wished to write an article upholding the doctrine of "punctuated equilibrium," and wondered if we would publish it. I wrote back saying we would be happy to do so, and reserved my right to respond, in turn, to him. I thought this might be the beginning of an intelligent communiqué. However, several months passed and no article. Then one day I received a letter indicating he had not had time (and could not foresee having the time) to pen an argument as promised. Instead he wrote the following insulting remark:

> As a joke I think articles knocking evolution are fine, but I think taking non-Darwinian explanations for the way things are is way out of touch with reality. It goes beyond the simplistic "The Bible is right" type of crap that we have come to expect from people with a certain set of short circuits in their brains.

This writer also enclosed a copy of a 4x5-inch pamphlet produced by the well-known atheist publishing house See Sharp Press entitled: "What God Has Revealed to Man." The cover of this pamphlet promised to reveal "The genuine Word of God, As Revealed by the World's Holy Men and the World's Holy Books." Inside were six blank pages. Very cute. And this writer had the nerve to use the word "simplistic" with regard to anti-Darwinian arguments?

My article was not at all tongue-in-cheek; neither did it claim that the Bible is right. The reader merely projected this assumption upon it. As a critical response to a well argued anti-Darwinian thesis, this atheist pamphlet was truly a sophomoric joke. I am very familiar with atheist arguments, having lived in that camp for over twenty years, and having come from that camp only recently. Let me say that these are generally the most shallow rationalizations on the face of the planet, and the anger coming out of them is serious (the word "short circuit" comes to mind).

Throughout my years of involvement in the conspiracy milieu, I have noted that people quite often use the term "simplistic" with regard to ideas and theories they haven't taken the time to study. After saying he was going to work on a response, this writer instead urged *me* to read Darwin's *Origin of Species* in its entirety! Therein, I suppose, the answer to my silent prayer for a brain would be found. My feeling is that I've already read that book, and processed it, and did *my* research, and it was *his* turn to do some reading, since he's the one who had a problem with my worldview in the first place. But, instead of doing *his* homework, he simply urged me to read *his* Bible, which apparently says it all for him! Open mouth, insert foot.

This insulting response made me fully realize that Christians are treated as though they are ignoramuses. In our years of publishing *Paranoia* magazine, my partner and I have published many articles written from a Christian perspective, while other magazines in the conspiracy genre, as well as other mainstream and independent publications, usually opt *not* to do so. What this person had not realized was that I am not a Christian, nor am I arguing from any religious viewpoint. I am simply addressing evolution on its own merits, as a thinking person. I am merely asserting that Darwinian theory cannot tell us anything about the genesis of the humanoid form at large in the Cosmos. It is unnecessary to insert another theory in its place. Is it better to stand upon a foundation of lies, or upon a bed of questions?

To utter the blasphemy that Darwin was wrong is comparable to Galileo's house arrest for stating that the Sun was the center of the Solar System. According to a strict Darwinian interpretation of evolutionary processes, allowing any outward teleological factor into its analysis—life force, Gaian archetype, or morphogenetic fields—would be conceding to the supernatural—the vital, the invisible, the external—and would spell the downfall of science. But perhaps our scientists have got it wrong. We have got to embark in a serious consideration of alternative possibilities.

It is suggested in this book that there may be a guided hierarchy of creation ongoing in the Universe. Various human genesis theories may be combined in an overall concept of material creation within this hierarchy. A vast body of theory comprising The Extraterrestrial Hypothesis, as discussed in the next few chapters, is just one of these alternative possibilities, and may simply be part of a universal creation dynamic, or may even be a localized aberration from the "normal" way of doing things. The following chapters provide an overview of a hierarchy of human creation culled from cross-cultural aboriginal sources.

Chapter Six

Galactic Surveys:

The Oral and Written Tradition of the Sirius System

For every perceivable phenomena devise at least six explanations that indeed explain the phenomena. There are probably sixty, but if you devise six, this will sensitize you to the complexity of the Universe.

Paula Underwood
"Who Are the Human Beings?"
When Cosmic Cultures Meet, Conference

The Space Travel Argument outlined earlier concludes that space-travelling ETI apparently do not exist, and that it is very likely that we are the only intelligent species now existing in our Galaxy. However, Astronomer Tom van Flandern has noted a certain erroneous "probability argument" popular among his scientific colleagues. Writing in *The Anomalist*:

it is often asserted that the probability of extraterrestrial intelligence visiting the solar system is *extremely small*. But that is not a *known scientific fact*. In truth, that probability is *unknown*, which is quite a different matter from being small. For all we know, it may be the case that nearly every terrestrial-type planet in our galaxy has already been visited by extraterrestrial intelligences, making the probability high, not small.

Van Flandern states succinctly the crux of a growing problem in scientific research in his article entitled "Betting on the Mars Face" published in *The Anomalist 2*. He writes that research on "alternative theories" are the first to be cut from tight funding budgets and that such proposals, if not presented within the confines of the mainstream paradigm, are inevitably denied funding. This explains why mainstream science journals are filled with papers supporting mainstream models, and explains also why alternative theories and those espousing them become marginalized. They are marginalized by the economics of the status quo scientific paradigm, one of which, as expressed by Van Flandern, is "that there has been no extraterrestrial intervention in the origin and development of life within the solar system, all of which is on Earth."

The recently discovered "anomaly" termed the "Face on Mars" is one very strong argument on the side of a new paradigm now emerging. The new view suggests that mankind is not alone in the Universe, and further that *Homo sapiens* did not "evolve," *at least not without possible genetic intervention* along with cultural indoctrination by one or more groups of extraterrestrial travelers.

In an internet update dated 7/18/98 following the Mars Global Surveyor mission, Van Flandern wrote that: "based on the best available high-resolution, contrast-enhanced Mars Global Surveyor image and the best old Viking images, the 'Face'mesa contains regularity, angularity, symmetry, and the fulfillment of a priori predictions based on the artificiality hypothesis, such as the appearance of nostrils in the nose, mouth shaping just under the nose, an eyebrow over the eye socket, a pupil in the eye socket, a separated vertical enclosure of the whole mesa with near perfect symmetry and corners, a crest over the headpiece, and the almost complete absence of extraneous or non-contributing features." (*www.metaresearch.org/announce/on-improbable-claims.htm*)

It might be advisable under the circumstances to utilize "The Rule of Six." This Rule states that "For every perceivable phenomena devise at least six explanations that indeed explain the phenomena. There are probably sixty, but if you devise six, this will sensitize you to the complexity of the Universe, the variability of perception. It will prevent you from fixing on the first plausible explanation as the Truth."

With regard to culture, German philosopher and Nobel prize winner, Friedrich August von Hayek, has pointed out that the system of behavioral rules we call *culture* originally "in all likelihood contained more 'intelligence' than did man's thought about his environment." In *The Origins of Life,* evolutionary philosopher H.V. Ditfurth suggests that mankind's "brain is capable of accepting culture, but not of devising it" and that culture is a kind of "knowledge without brains." In von Hayek's view, "the intelligence of *cultural systems* has been *far superior to the intelligence of individual brains for much of human his - tory.*" He asserts that it is "misleading to consider the individual brain or the individual mind as the keystone in the hierarchy of complex structures produced by evolution."

To state this a different way, cultural adaptation has taken over for environmental adaptation *since the earliest times*. No anthropologist would argue that there is an enigmatic collectiveness and beyondness in the phenomenon of culture. Ditfurth notes that "people began to describe their reality by means of art thousands of years before their individually accumulated knowledge enabled them to realize that language alone couldn't handle the job." (Ditfurth, 192) If the sudden emergence of the earliest cultural systems contain features that are *beyond man's innate intelligence to devise*, where did it come from? It is almost counterintuitive to posit that this distinct human quality originated in the *indi - vidual brain or individual mind* of one evolving ape, yet evolutionists go through all sorts of contortionism to make this explanation fit.

The Dogon Tribe

Archaeologists have great difficulty explaining the similarities between the cultures of Sumer and Egypt, and have supposed that a common origin for these two cultures must exist. Egyptologist Wallis Budge was convinced that Babylonia and early Egypt derived their cultures from the same "exceedingly ancient" source. Budge has suggested that the profound similarities between the religious systems of these two cultures could not be the result of borrowing, and argues that this "company of primeval gods" is quite different from the Semitic gods which arose later in Babylonia and Assyria.

According to Robert Temple in his book *The Sirius Mystery*, the writings of the ancient Egyptians exhibit the same Sirius tradition found in the oral tradition of the present Dogon tribe of Mali, located in sub-Saharan Africa near the Ivory Coast. It is believed that the Dogon are most probably related to Lemnian Greeks, who claim descent from "the Argonauts." The Sirius tradition is traceable from pre-dynastic Egypt (prior to 3,200 B.C.) to the Greeks who took it to Libya, and moved west and south finally reaching the Niger River and intermarrying with the people of Mali.

The Dogon sacred mystery tradition includes extensive scientific knowledge of the star systems of Sirius A and Sirius B (and possibly a presently unknown system 'Sirius C'). According to Temple, the Dogon present a theory of Sirius B which fits all known facts about it, and possibly more facts which are simply unknown at this time. But how could pre-dynastic peoples have had such extensive knowledge of planets we have only recently been able to see with the use of modern telescopes? An article published in 1950 by French anthropologists entitled "A Sudanese Sirius System" reports on four tribes in the French Sudan, the Dogon and three others, which share a secret religious system concerning the star system Sirius. The religion incorporates scientific knowledge about the star system which *should* be impossible for primitive peoples to know.

The most important star to the Dogon tribe is Sirius B, a tiny star which orbits the bright "dog star" Sirius. The Dogon say that Sirius B is "invisible." The star Sirius B is, in fact, totally invisible except through a very powerful telescope. How do they know it is there? The Dogon call this star *po tolo*. The cereal grain which they call *po* is the smallest grain known to the Dogon; it is extremely tiny. The word *tolo* means star. Thus, the Dogon's *po tolo* means *tiny star*. In both oral reports and in their drawings, the Dogon expressly describe the orbit of Sirius B around Sirius A as egg-shaped, or elliptical. It was Johannes Kepler who showed us that heavenly bodies do not move in perfectly circular paths; how do the Dogon know this?

Robert Temple asserts that comparison of the Dogon star maps with accurately scaled modern astronomical diagrams of the Sirius system demonstrates a striking similarity. The fact is, he writes, "the Dogon have an accurate general knowledge of the most unobvious and subtle principles of the orbiting of Sirius B around Sirius A." The Dogon are also aware of the 50 year orbital period of this invisible star, as well as the fact that it rotates on its axis. They also know that the turning of the Earth on its axis is what makes the sky seem to turn around. Thus, the Dogon have, for a very long time, been "free from the illusions of our European ancestors."

The Dogon say that Sirius B is the origin of all things; it is the "egg of the world." It lies at the center of all things and without its movement no other star could hold its course. In particular, it governs the position of Sirius A, the unruly star, by encompassing it with its trajectory. In the Dogon tradition, Sirius B gives birth to everything that exists, visible or invisible. It is made up of four basic elements: air, fire and water. The fourth element is metal. Yet, the Dogon describe Sirius B as infinitely tiny. It is known that Sirius B is, in fact, a white dwarf, the tiniest form of visible star in the Universe. What is even more profoundly astute is the Dogon's knowledge that "the star which is the smallest thing in the sky is also the heaviest." The Dogon explain that this star is made of a metal called *sagala* which is a material that does not exist on Earth. How do the Dogon know that Sirius B is, in fact, made of super-dense material which exists nowhere on Earth?

The Dogon people are also familiar with numerous other heavenly bodies. They know of the four major Galilean moons of Jupiter. They know that Saturn has a permanent halo around it. The Dogon are aware that "an infinite number of stars and spiraling worlds exist" and that all types of creatures live on other "Earths." In addition, they say that the Nommo, the people who came to Earth in their spaceships, will come back again when their 'star' appears in the heavens as "testament to the Nommo's resurrection." They state that when the Nommo originally landed on Earth, he "crushed the Fox, thus marking his future domination over the Earth." Robert Temple takes this to mean that "man's brutish nature was subdued in the distant past." They refer to Nommo as the monitor of the Universe.

In addition, a third star in the Sirius system is described as being "four times as light (in weight)" and travels along "a greater trajectory in the same direction and in the same time" as Sirius B. (Temple, 24) The third star in the system also has a body which orbits it. We have not as yet discovered the third star, but the Dogon say it will appear. And, of course, just like the heavenly bodies which revolve around Sirius A, the secret traditions of the Dogon revolve around aboriginal tales and personages connected with these astronomical bodies. These tales and personages are much the same as the Sumerian written tradition.

History or Mythology?

Joseph Campbell was a Jesuit-trained scholar who gathered myths and garnered information from varied disciplines to construct a sort of "science of the unconscious," or, what Neil Freer describes as a "sociobiology of revelation." Writing in the book *Of Heaven and Earth,* Freer concludes that Campbell was never able to interpret the "gods" as historical figures. He was only capable of psychoanalyzing myths as projections of mankind's mind as it attempted to make sense of and order reality. Campbell could only hold, Freer writes, "that their source had to lie in the psychology of the minds of men, even going so far as to say that there was madness in the claims of the god-kings of Egypt." Freer suggests that Campbell was trying to make sense out of an historical enigma "full of deep contradictions and mind boggling facts and seeing madness by psychoanalyzing the mess." In other words, if you're looking for madness, psychology is certainly the tool to use.

According to Freer, Campbell identified Sumeria as the "primary mythogenic zone," and as a prime source of the origination of universal mythic themes, but nonetheless continued to explain the gods as "mythic archetypes," applying theories of psychological projection. Freer asks, "how else can you explain something that you are convinced is unreal?" Campbell continued to refer to the amazing cross-cultural similarities in the genealogies of the gods as a process of "diffusion," which is really a non-explanation. Princeton psychologist, Julian Jaynes, also thought the ancients were crazy folks, suggesting that, since people before 1250 B.C. claimed to be in some form of communication with the "gods," the people of this era must have been schizophrenics. Freer concludes that the "Jaynes-Campbell Syndrome" is "the most extreme explanation one can advance if one starts from the unquestioning assumption that the gods must have been unreal."

The Freudian/materialist paradigm should be construed as part of a pervasive philosophical mindset that is the antithesis of arcane knowledge about human origins. As such it stands in the way of the Revelation of the Method: full recognition of the occult nature of the ancient past.

The Ancient Gods: Planets or People?

The polarity of thought in the modern struggle to elucidate human origins is most evident in the question of how to interpret ancient myths. In his book, *Hamlet's Mill*, Giorgio de Santillana describes the planetary mythos and its "fundamental archaic design" in fascinating detail. He writes (p. 177):

> The real actors on the stage of the Universe are very few, if their adventures are many. The most "ancient treasure"—in Aristotle's words—that was left to us by our predecessors was the idea that the gods are really stars, and that there are no others. The forces reside in the starry heavens, and all the stories, characters and adventures narrated by mythology concentrate on the active powers among the stars, who are the planets. A prodigious assignment it may seem for those few planets to account for all those stories and also to run the affairs of the whole Universe. What, abstractly, might be for modern men the various motions of those pointers over the dial, became, in times without writing, where all was entrusted to images and memory, the Great Game played over the aeons, a never-ending tale of positions and relations, starting from an assigned Time Zero, a complex web of encounters, drama, mating and conflict.

The imagery contained in creation myths, de Santillana argues, are attributes of the planetary systems, movements, and positions, and the related astrological signs of the zodiac, overlaid with human, animal, and terrestrial-based descriptions. Yet, we must wonder, how did the ancients arrive at their enormously extensive knowledge of these distant bodies, which is evident in their mythologies? We should certainly not presume that they were less intelligent than we are, but the real question is, via what technology were they able to know the minutest of details regarding our solar system? It is true that the ancients had vast knowledge and also had remarkable technologies, but here we are talking about a very ancient "aboriginal" mythogenic source that is supposed to be pre-civilization.

The fundamental difference in the translation and understanding of ancient mythologies boils down to a tiff between archaic/astrological/extraterrestrial vs. biological/natural/terrestrial holdings. Some scholars believe aboriginal creation myths described real earthly events involving cosmic catastrophes and the people who were actually alive to witness them, and who told of 'gods' who walked among them, before and after, who established and re-established the ancient civilizations. Others are of the opinion that these stories are a "psychology" and not a "history;" that they are projections of the aboriginal mind onto the panorama of the skies, an archaic "mind reel." It is interesting that humans continue to play out this astrological/extraterrestrial "War of the Worlds" within a terrestrial "War of Words" (psychology/biology); truly an enormous Glass Bead Game.

As de Santillana writes, it is indeed a "prodigious assignment" for a few planets to account for all of these stories; as he describes it: "a never-ending tale of positions and relations, starting from an assigned Time Zero." Indeed, he is correct to assert, the actors in this universal saga are few, but the details of their exploits are manifold. But, we must also question, who devised this "assigned Time Zero"? Who assigned a beginning to the zodiac? Where would "archaic" man get such information?

It would appear that there is indeed a never-ending drama which repeats itself cross-culturally and through time in the guise of various cultural motifs. We are still playing out that drama with the modern origins story of evolution (terrestrial/biology) vs. creation (extraterrestrial/astrological). Where truly lies the leap of faith required to bridge the gap between the ancient world and the modern world; under which rock will we find our biological, psychological, and astrological origins?

In her prolific web site writings, Acharya S., author of *The Christ Conspiracy: The Greatest Story Ever Sold*, writes that we must be able to discern between the *gods* and the *sky people* mentioned by the ancients. The gods were the seven planets, she writes, but the sky people may have been real historical figures. Acharya writes that *some* of these sky people may have been what we refer to as *aliens*, since ancient texts indicate that "advanced people appeared around the world to reestablish civilization after the various cataclysms." In doing so, these "priests also reintroduced the gods of the mythos." As time went on, the teacher became associated with the god, and the mythos became entwined with the *history* of the teacher, who was merely a representative of the god. Yet, if they were simply human "priests," where did they hide during the great cataclysms and how did they make worldwide appearances afterwards?

Acharya argues that the Sumero-Babylonians identified their gods as planets, including the sun and the Moon. Specifically, the Anunnaki were called "the fates," and were the "seven nether spheres." She argues that they were not persons, nor were they aliens. She argues that the "personification of the planetary bodies" and "vulgarization of the celestial mythos and ritual" can be blamed for a resulting loss of knowledge. This body of knowledge, she writes, is traceable through legends to 70,000 years ago, and perhaps much further back. She believes that "to reduce this glory to a band of aliens and/or humans is silly and deplorable, as it robs the ancients of intelligence and wisdom." She insists there is no need for "absurd sci-fi explanations," such as "a group of bizarre aliens terrorizing cavemen."

William Bramley counters this viewpoint in *The Gods of Eden*. He asserts that the "gods" of Egypt's early period were described in terms of being "literal flesh and blood creatures" with the same needs for food and shelter as human beings, and who traveled in flying "boats." The homes of these gods were supplied with human servants who later became the first priests, and were initiated into the Brotherhood Mystery Schools. The derailment of spiritual knowledge in Egypt, Bramley claims, was caused by the corruption of the Brotherhood of the Snake, or the Mystery Schools, which, he maintains, was a tool of the extraterrestrial "Custodial" society.

Bramley asserts that this group of space travelers introduced the world's religions to the people as a means of social control. He aptly illustrates that, via a network of Brotherhood organizations, wars were continuously generated by a Custodial [alien] society. The Custodial Brotherhood network was designed to obfuscate and eventually obliterate spiritual knowledge. For instance, he writes:

> Hindu writings indicate that people of diverse races and personalities made up the Custodial [alien] society ... by the time of the Aryan invasion, the oppressive ones were clearly the dominant ones. This was evident in the social system imposed on India by the Aryans. That system was unmistakenly designed to create human spiritual bondage. As elsewhere, this bondage was partially accomplished by giving spiritual truths a false twist.

Nobody who has fully explored the UFO enigma could deny that any of its peculiar scenarios are bizarre and absurd. Yet, the fact that the ET hypothesis may be construed by the modern mind as an "absurd sci-fi explanation" does not necessarily make it an incorrect interpretation of an historical "mythos." We may not like the implications, we might think it's weird, but we cannot be so sure that the ancient 'gods' weren't extraterrestrial people.

There is strong evidence that actual cosmic collisions and near collisions have occurred in our own planetary system, and the memory of these events was perhaps solidified in stories, and played out over and over again, with "biological" details added. The biological details are what we are arguing about here. Were the 'gods' simply advanced people, terrestrial-based 'priests' of a higher caste or class, or were they extraterrestrials who landed in spaceships and eventually conferred kingship upon certain humans? At what point did this occur and what relationship, if any, did these "advanced" persons have to the ancient gods represented by the planetary system of the zodiac? When does pre-history become history, and when does archaic man become modern man? If, as Velikovsky has suggested, ancient man emerged from each cataclysm thinking he was at Time Zero, how do we even begin to contemplate the idea of "In the Beginning"? Exclusion of any of these considerations causes us to rush to premature judgment.

Organic Time and Cultural Time

The cosmology of remote antiquity, an intertwined astrology and astronomy, was an immensely sophisticated science, writes Giorgio de Santillana in *Hamlet's Mill*. Further, he asserts, this ancient science was not founded on any system which could be considered a basis for teaching it. De Santillana posits that this science "existed before systems could be thought of," and, he suggests, it was "spontaneously generated." The implications of this observation upon Darwinian evolution and the emergence of culture, as well as the meaning of the vast mythology which represented this cosmology, are worth considering within the present context.

Just as we talk of the "Newtonian period" of two centuries ago, writes de Santillana, our period may some day be called the "Darwinian period." He writes: "The simple idea of evolution, which it is no longer thought necessary to examine, spreads like a tent over all those ages that lead from primitivism into civilization. Gradually, we are told, step by step, men produced the arts and crafts, this and that, until they emerged into the light of history." He adds (p. 71):

> In later centuries historians may declare all of us insane, because this incredible blunder was not detected at once and was not refuted with adequate determination. Mistaking cultural history for a process of gradual evolution, we have deprived ourselves of every reasonable insight into the nature of culture. … [For,] what are our natural principles but principles of custom.

This seeming and imposed "gradualness" is only gradual due to certain cultural biases. The gradualness of cultural acclimation, we must consider, is itself a cultural artifact of the present "Darwinian period." As de Santillana writes, the words "gradually" and "step by step," are aimed at "covering an ignorance which is both vast and surprising." If we wished to know more about these supposed "steps," there would be no answer. We could only be "lulled, overwhelmed and stupefied by the gradualness of it all, which is at best a platitude, only good for pacifying the mind, since no one is willing to imagine that civilization appeared in a thunderclap."

But if civilization appeared in a thunderclap, from whence came this ancient science, which appears to have "no system for teaching it"? From whence came a "spontaneously generated" science with no underlying system or paradigm? Myth-making as strictly preternatural observation is an intriguing theory, and it's important not to assume that ancient peoples had less intelligence than we do. Perhaps they had more intelligence, and a real "connection" to the Universe which we have since lost. But pointing to some vague "archaic" and "spontaneous" cosmic connection is akin to saying that "step by step" we gained something and "step by step" we lost it again. Is this in fact what happened?

It is impossible to subtract catastrophic theory from the seemingly incremental achievements of our ancient ancestors. This step-by-step enigma is caught up in catastrophism, since the overwhelming feeling is that those steps are more like a "cha-cha": three steps forward, three steps back. Perhaps the idea of incremental increases in the complexity of cultural packages can also be described as a repetitive loss/gain syndrome caused by various catastrophic events. As de Santillana writes, "The lazy word 'evolution' has blinded us to the real complexities of the past. That key term 'gradualness' should be understood to apply to a vastly different time scale than that considered by the history of mankind." As he also points out, "organic evolution ceased before the time when history, or even prehistory, began. We are on another time scale. This is no longer nature acting on man, but man on nature. People like to think of a constancy of laws which apply to us. But man is a law unto himself."

Indeed, "something" happened on the way to history. We can't actually describe what we're missing, but it appears we "must have known" something, since, the mainstream paradigm argues, the mythology/astronomy cultural package was "spontaneously generated" from the "archaic" mind. Suddenly, nature doesn't act on man. There's no more gradualness. The terminology of "gradualness" veils the past and blurs the line between history and pre-history. We are on a different "time scale," only because somewhere on the way to history we became profoundly aware of our humanity; we became conscious beings. And we've been in shock ever since, because we woke up and didn't know who we were.

In his conclusions at the end of *Hamlet's Mill*, de Santillana wonders what the original "universe of discourse" could have been, as he calls it: the "insensate scattering of dismembered and disjointed languages of the remote past," from which, by some *"stroke of luck*, scientific man was born." He suggests that archaic man must have had the "capacity for attention, for singling out certain unattainable objects in the Universe." He adds:

> There were some men, surely exceptional men, who saw that wondrous points of light on high in the dark could be counted, tracked, and called by name. The innate knowledge that guides even migratory birds could have led them to realize that the skies tell the glory of God, and then to conclude that the secret of Being lay displaced before their eyes.

De Santillana posits that these strange "innate" and apparently "lucky" ideas born in the archaic mind were the beginnings of intellect, and, over time, he surmises, these ideas became a universal language which covered the globe. This overarching common language ignored local beliefs and cults, he asserts, and grew to encompass "numbers, motions, measures, overall frames, schemes, on the structure of numbers, on geometry." It did all of this, he asserts, even though "its inventors had no experiences to share with each other except the events of their daily lives and no imagery by which to communicate except their observations of natural lawfulness." How can we call this anything other than a "stroke of luck"?

The archaic past, writes de Santillana, was based on a "high culture" of the arts and sciences, whose ideas were at least as complicated as ours. It was in essence the first Technological Revolution. This ancient world view wove together geography and a "science of heaven," which included the fate of mankind's soul. The ancient concept of time was cyclical, and it derived from the revolutions of the stars and planets. Also important was the concept of numbers as "the secret of things." He writes, "those unknown geniuses set modern thought on its way, foreshortened its evolution." Cosmological Time, he explains, was not "an empty container," but was potent enough to control events of this world. De Santillana proposes that the transcendent structure of the archaic Universe supplied a "foundation to reality that all of modern physics cannot achieve." He writes:

> The strange hologram of archaic cosmology must have existed as a conceived plan, achieved at least in certain minds, even as late as the Sumerian period when writing was still a jealously guarded monopoly of the scribal class.

Were the gods "planets" or were they "people"? Who originated the celestial circle and divided it into 12 zodiacal houses? As Sitchin writes in *The Cosmic Code*, underlying this division of the heavens is a highly sophisticated astronomy which is so advanced that humans could not have developed it at the time it arose. The zodiacal circle and the astronomy it represented required a knowledge of the Precession of the Equinoxes, an astronomical fact which could not have been known without advanced observational tools. Sitchin suggests that the "impossible knowledge" of the Sumerians was imparted to them by "Anunnaki" space travelers, in a series of texts forming the basis of ancient science and religion, and was translated into many languages, including Greek and biblical Hebrew. All ancient mythologies stem from this original knowledge imparted to earthlings by space faring ETI, who named this planet ERIDU, meaning "Home away from home."

Sitchin's argument in *The Cosmic Code* is that the Anunnaki used their detailed sky charts to group the stars into constellations, and then honored their leaders by naming the constellations after them. Thus, the Sumerian god EA, ("Whose Home is Water") was honored by the zodiacal signs of Aquarius and Pisces, and the "priests who oversaw his worship" were dressed as Fishmen. Enlil, the strong-headed one, was honored by the sign of Taurus. Ninmah was Virgo. The warrior Ninurta was Sagittarius. Over time, Sitchin explains, as second and third generation Anunnaki 'gods' joined the scene on Earth, "all the twelve zodiacal constellations were assigned to Anunnaki counterparts." He argues, "not men, but the gods, devised the zodiac."

De Santillana writes that this complex astronomical system took shape by about 4,000 BC, yet, this cosmology is substantially older than this, since it predates the deluge which has been dated at approximately 9,500 BC. The tradition then suffered a loss in the Greek Middle Ages. Plato and Aristotle thought of these ancestors as "the men close to the Gods." At about A.D. 60, Plutarch wondered why "oracles had ceased to give answers." When and why did the "archaic world," with its connection to the gods, come to an end? Could it be because the gods departed? As the Emperor Tiberius decided, another world age must have passed away, and took with it the gods who belonged to it. Recall also the *Virgin of the World,* which describes Hermes, a representative of a race who taught mankind the arts of civilization, and then, "with charge unto his kinsmen of the Gods to keep sure watch, he mounted to the Stars."

We are indeed the lost civilization. We assume we must be missing something, our "animal" nature, our "archaic" connection to the Universe. But, it's something we can't talk about in more specific terms. Just as the terminology of "gradualness" blurs the line between history and pre-history, the terminology of "archaic" man serves to blur the transition between something lost and something gained, or between such contrivances as "organic time" and "cultural time." Where is this "something lost" which apparently included a self-teaching system? And how can we effectively argue that within this mysterious "something lost" lies the answer to whether mythological personages were planets or whether they were people (i.e. psychological projections vs. historical realism)?

The archaic world can be thought of as a world with one language. Could it be that what we lost is the mother tongue, and with it the universal understanding of the exogamous origin of humankind? It makes perfect sense that what was lost on the way to Babylon was the source of this "hologram" of the gods, who departed the Earth leaving the meaning and origin of these ancient stories dangling in mid-air.

Indeed, as Neil Freer says in his essay entitled "From Godspell to God Games," the Nefilim (Sumerian gods) inexplicably "phased off the colony planet;" they simply left "without closing the laboratory door," around 1250 B.C. The foremen-kings, he states, are suddenly depicted in stone carvings pointing to the master's empty chair in utter dismay. The engravings state: "What do I do when my master is no longer here to instruct me ... what shall I tell the people." Grieving, they looked to the sky for the return of the gods, in which posture we still stand today.

Time Capsules

Rather than take a well-trodden path, the path which conforms with, in Charles Fort's words, an "attempt to hold out for isolation of this earth," an alternative analysis needs to begin with a different premise. Rather than, as Fort so poignantly states, "ignoring externality to the greatest degree possible, the notion of things dropping in upon this Earth from externality," we need to ask, IF historic contact from *externalia* actually occurred, *how* might the memory of ETI contact be preserved and passed on? We also need to realize that such written and oral "time capsules" would be preserved in various cultural forms and in a certain historical context, and on top of that would be subject to the cultural biases and interpretations of the people who later retrieve them.

In light of facts which have come to fore, it is important to open these time capsules again without the cultural bias of isolationist doctrines like neo-Freudianism; not seeing them as elaborate internal projections imposed onto the outer world of man, but as real events which impinged upon him from externalia.

The question of how the memory of ETI contact might be preserved was addressed in a book entitled *Intelligent Life in the Universe*, which was co-authored by astronomer Carl Sagan and I.S. Shklovskii of the Soviet Academy of Science. Sagan writes that "there are no reliable reports of direct contact with an extraterrestrial civilization in the last few centuries." However, extrapolating from traditional contact stories, he writes that contact scenarios are often "encumbered with some degree of fanciful embellishment, due simply to the views prevailing at the time of contact." He also explains that "the extent to which subsequent variation and embellishment alters the basic fabric of the account varies with time and circumstance." He writes that some oral renditions contain sufficient information for later reconstruction of the true nature of the encounter, although many incidents [may be] disguised in a mythological framework."

Sagan specifically discusses the Dogon contact scenario, concluding that "such legends and myths, handed down by illiterate people from generation to generation, are in general of great historical value." In this chapter of his book, which is provocatively titled "Possible Consequences of Direct Contact," Sagan discusses the creatures credited with the founding of the Sumerian civilization. They are described as "amphibious" and as being "happier if they could go back to the sea at night." These entities were "superhuman in knowledge and length of life." They eventually returned in a ship 'to the gods' carrying with them "representatives of the fauna of the Earth."

Sampling the Third Planet of a G Dwarf Star

Sagan seems to take the Dogon story somewhat seriously. He goes on to postulate that if Earth had in fact been visited in the distant past, "matters of evolution, while difficult for us to reconstruct from a distance of millions of years, would have been much clearer to an [advanced] technical civilization." He also surmises that if such an advanced culture visited the Earth just once, then they might "visit every hundred thousand years or so to see if anything of interest was happening lately." He speculates that "some 25 million years ago, a Galactic survey ship on a routine visit to the third planet of a relatively common G dwarf star [our Sun] may have noted an interesting and promising evolutionary development [us]."

Sagan surmises that the news of intelligent life having evolved on Earth would travel the stars and "the rate of sampling of our planet should have increased, perhaps to once every 10,000 years." Sagan further posits that "the development of social structure, art, religion, and elementary technical skills, should have increased the contact still further." He deduces that "if the interval between sampling is only several thousand years, there is a possibility that contact with an extraterrestrial civilization has occurred within historical times."

Sagan's speculations are interesting simply because they were made by a person in the scientific establishment. However, one must keep in mind that his point of view is typically anthropocentric, allowing for a strictly linear development beginning with the gradual, accidental and *local* human *evolution*, followed by an unexplainable burst of cultural evolution, followed by ETI "sampling." The theoretical civilization of which he speaks is merely 'taking notes'. They seem to be cosmic anthropologists, fascinated with the pristine local culture of evolving apes on a beautiful blue oasis. Each time the galactic grad students visit, every 100,000 years or so, they take more notes. Maybe they stop by whenever possible every 10,000 years so they can take more notes. After all, it is a "Galactic survey."

The problem with this scenario is that it ignores the factual evidence sitting directly under our noses, preferring to speculate about theoretical survey ships and sampling rates. In fact, in temples and palaces, in centers of commerce and administration, and in state and private archives and libraries all over the ancient Near East, clay tablets by the tens of thousands etched by ancient scribes record a detailed history of ETI contact with Earth. As Zecharia Sitchin writes in *Wars of Gods and Men*: "this meticulous lot used monuments, artifacts, foundation stones, bricks, utensils, weapons of any conceivable material, as inviting slates on which to write down names and record events." Many tablets are identified as copies of older tablets which had been written in the "olden language." What more *evidence* do we need that this is historical documentation and not myth-making?

This written history illustrates that the course of human development occurred along drastically different lines than Sagan, and others who entertain the possibility, might consider. Rather than gradual, accidental and local human evolution being followed by cultural evolution and ETI "sampling," we are actually informed by prolific Earthling note-takers that ETI visits came *first*, and that the emergence of humans was *purposeful, nonlocal, and rather abrupt*. In fact, we are even told that the so-called "rate of sampling" is closer to 3,600 years, and that at least one phase of it began with a Galactic "survey" about 442,000 years ago. How's that for details?

Galactic Survey: The Anunnaki of Nibiru

According to biblical scholar and ancient historian, Zecharia Sitchin, not only has Earth been visited by at least one ETI group, but humans and Earth civilizations are the direct result of genetic manipulation and cultural indoctrination programs carried out by space travelers hailing from our own solar system. The reasons for these genetic and cultural programs, he explains, are outlined in the ancient Akkadian/Sumerian, Babylonian and even Biblical texts.

Archeologists agree that the remarkable features of this high civilization, which included advanced technologies, social organization, sciences, arts, education and law, appeared *abruptly* and without precedent in the area called Sumer in the Mesopotamian region. The archeological record is absent any pre-human features which might have led to the enormous cultural complexity of this ancient civilization.

For instance, by the third Millennium B.C., the Sumerians had developed advanced systems of writing, printing, agriculture, mathematics, astronomy, calendrical system, trading, banking, shipping and complex metallurgy, which included the transportation of metals and other building materials over vast distances. The Sumerians had intricately woven fabrics and textiles, clothing, and jewelry, as well as the world's first boats. Sitchin concludes in *The Twelfth Planet* that "the ability to carry out major construction work according to prepared architectural plans, to organize and feed a huge labor force, to flatten land and raise mounds, to mold bricks and transport stones, to bring rare metals and other materials from afar, to cast metal and shape utensils and ornaments—all clearly speak of a high civilization, already in full bloom in the 3rd Millennium BC."

In addition, the Sumerians had various types of land vehicles, such as carts and chariots, which were drawn by horses and oxen. The Sumerians were also using petroleum products for fuel, waterproofing, painting, cementing, road building, molding, chemistry and medicine circa 3,500 B.C.. The detailed architectural floor plans and instructions for building their enormous temples were said to have come from the 'gods' and required professional transcription. Along with these instructions from the gods was given a measuring rod and a roll of string for the job. The temples, Sitchin clarifies, were but "the tip of the iceberg of the scope and richness of the material achievements" of this high civilization in ancient Sumer.

Sumerian writings indicate that various building materials for these high-rise temples in Mesopotamia, such as gold from Africa, silver from the Taurus Mountains, and cedar from Lebanon, were brought by the Sumerians in ships. An Akkadian dictionary of the Sumerian language, for instance, contains a section on shipping which lists 105 Sumerian terms for various types of ships according to their size, destination, or purpose (for instance, ships for cargo, for passengers, or for the exclusive use of certain gods), and another 69 terms relating to the construction and manning of ships. Sitchin writes in *The Twelfth Planet*: "only a long seafaring tradition could have produced such specialized vessels and technical terminology."

Sumerian legal and medical systems provide an indication of the age of the Sumerian culture. The reform decree of the *ensi* (leader) named Urukagina at about 2600 B.C., which Sitchin describes as "man's first social reform based on a sense of freedom, equality and justice..." specifically reaffirms social protections, such as the rights of the blind, the poor, widowed, orphaned, and divorced. The King emphasized that the gods had decreed a return to an *earlier* tradition. Sitchin also reports that a tablet from Ur dated 5,000 B.C., a medical text dealing with diagnosis and prescriptions, describes common medical practices in use by the Sumerians which apparently followed sound scientific principles. Thus, while the exact age of the Sumerian culture is unknown, it is surmised to be well beyond 5,000 B.C., since records implicate a social order that came before the phase of occupation currently under study.

Although the previous information is assumed to be an historical record of the earliest Earth civilization, the more "bizarre" attributes of the story which pertain to the cosmology of the Sumerians, although derived from the same source, are considered to be its *mythology*. It would appear that assumptions of *fact* and *fiction* are arbitrarily assigned to this information with regard to what is 'believable' and what is 'far out'. It is important to realize though, that the ETI contact story of the Sumerian culture has been passed on to us in two ways. It has survived to this day as the oral tradition of the Dogon people, who have maintained minimal contact with the outside world, and it was passed on via the prolific writings of the Sumerian civilization. In addition, various world stories, including the well known Greek sources, exhibit some of the same stories and personages.

While some might take this contact account to be a *mythology*, since it contains attributes which might be construed as Sagan's "mythological embellishment," Sitchin believes these written accounts, which are expressed as a central tenet rather than as a theatrical sidebar, are the true pre-history of Earth civilization and the cosmic birth of humankind. In all of these writings, Sitchin stresses in *Wars of Gods and Men*, "the same facts emerge as an unshakable tenet of the Sumerians and the peoples that followed them ... [that] the "Righteous Ones of the Rocketships," the beings the Greeks began to call the 'gods', had come to Earth from their own planet." He notes that such statements were not made lightly but, rather, "in text after text, whenever the starting point was recalled, it was always this: 432,000 years before the deluge, the DIN.GIR came down to Earth from their own planet."

We must ask ourselves why we remain so incredulous about this idea. Even though we are "aware" that we live in an immense Universe, and our culture considers itself ready for galactic exploration and biological dispersion, we maintain an oddly Earth-centric and insulated posture?

Visits From Outer Space

In 1940, while working on his manuscript *Worlds in Collision*, Immanuel Velikovsky claimed that a passage in the book of Genesis caused him to wonder whether it outlined "a visit from space." The passage which caused his wonder was the one which told of how the sons of God came to the daughters of men. Because the story sounded so fantastic, he decided not to publish anything on the subject. When UFOs began to be seen in the 1950s, Velikovsky noted that the view that these were visitors from other planets "does not find any credence with me."

Unable, for the most part, to translate world cataclysms within the panorama of space traveling ETI, Velikovsky nonetheless made some assertions to the effect that such ideas should not be automatically excluded. He wrote, "If we are today on the eve of interplanetary travel, we must not declare as absolutely impossible the thought that this Earth was visited, ages ago, by some people from another planet." Velikovsky wondered whether this peculiar passage in Genesis was a "literary relic dealing with a visit of intelligent beings from another planet." He also queried, "was this Earth alone populated by intelligent beings?"

Velikovsky then began making room in his theory for space travelers in connection with the deluge. He suggested: "It appears that the extraterrestrial visitors made their landing as if in advance knowledge of the impending catastrophe of the deluge. It could be that Jupiter and Saturn were approaching each other even closer on their orbits and that a disruption of one of them was expected." Velikovsky even allowed room for the notion that perhaps these visitors rested a while in paradise. He stated: "possibly many centuries, or even millennia, passed between the landing and the deluge." A mission could have been undertaken in order to study the conditions on Earth, he surmised. Or, perhaps, it was an escape from one of the catastrophes which preceded the deluge.

Furthermore, Velikovsky explained, if these people were fleeing some catastrophe, their ships may have made crash landings on Earth, some surviving and some not. He wrote that the Book of Enoch may describe such a group, "which was composed of males only, two hundred in number, under the leadership of one by the name of Shemhazai." In addition, Aggadic literature states that these "sons of Gods" tried to return home, but could not. Velikovsky guessed that these men would have been of gigantic stature. The great size of the visitors, he wrote, suggests that they came from a smaller planet with less gravitational influence than Saturn.

The Galactic Dragon Community

The true origin of the Anunnaki and other mythological deities is provided in detail by Mark Amaru Pinkham in *The Return of the Serpents of Wisdom*. Pinkham writes that the Serpents of Wisdom can be traced to the dawn of time when all that existed was an unlimited ocean of consciousness, a spiritual sea of androgynous, unmanifest Spirit. The Primal Serpent was the first tangible form assumed by the Spirit, he writes. Through this vehicle, God created the entire universe, using the triune powers of creation, preservation and destruction.

The hierarchy of creation included several Orders of Angels, headed by the Sacred Seven, or the seven Archangels. Pinkham writes: "At the beginning of time they assisted in the creation of the cosmos and today they rule over the Serpent's seven principles as they manifest within the physical universe."

After the Archangels in the natural hierarchy come the Seraphim, followed by their cousins the Cherubim. Under these angelic orders of the hierarchy come the 70 angelic orders overseeing different aspects of creation. According to Pinkham's research, below these angelic Serpent orders are "Orders of Extraterrestrial Serpents of Wisdom." These are physical serpent forms who inhabit other planets, star systems, and galaxies. These Intergalactic, Interstellar, and Interplanetary Serpents, Pinkham writes, travel freely throughout the physical universe. They travel by using their "immortal dragon bodies" and/or by advanced space traveling technology. (Does this sound like the "merkaba" discussed earlier?)

Pinkham writes that the Extraterrestrial Serpents travel to new or evolving galaxies and solar systems in order to assist "fledgling life forms" through their cultural development, and to help facilitate a "paradigm shift" when necessary. He writes that the Serpents of Wisdom have been coming to our solar system for millions of years for this particular purpose. During periodic visits, these ETs established bases in our solar system and built pyramidal structures in many planets, including Mars, Venus, and Earth.

The beginning of the present 104,000 year cycle, Pinkham explains, was the time when many ET Serpents came to Earth. He writes (p. 6): "The wisdom brought to Earth at the beginning of the cycle was calculated to assist in developing humankind throughout the cycle's entire duration, but especially during the major paradigm shift at its conclusion."

The Angelic and Extraterrestrial Serpents of Wisdom began arriving on Lemuria from "many corners of the cosmos," in order to participate in the creation of a divine paradise. Following their creation of the Lemurian landscape according to the blueprint of the Divine Mind, some of the creator angels, known in the Vedic tradition as the Devas, decided to remain on Earth as "protecting nature spirits and devas for the duration of the cycle." Also, many ET Serpents came to Lemuria to serve as teachers and priest kings for the fledgling human race. As Pinkham writes (p. 12):

> Collectively, they, and the Angelic Serpents, are mentioned in many creation myths worldwide as the two, four, or seven immortal Twin Sons of the Solar Spirit who arrived on Earth as creators and culture bearers at the beginning cycle of time.

More ET Serpents arrived from their Intergalactic posts to colonize the Dragon Land of the Atlantic: the lost continent of Atlantis. Pinkham notes that some of the Interstellar Serpents who settled in Atlantis hailed from the Pleiades star system. The symbol of the Pleiadian Serpents was the caduceus, the serpent and staff, which shows up in many alien abductions. According to Cherokee teachings, these "androgynous" Serpents are known as the Sacred Seven, and they came from the Pleiades, the Seat of the Divine Mind, in order to "instill within developing mankind the spark of individuated mind (the intellect and sense of separate self)." The Pleiadians mated with earth humans in Atlantis. Ancient Greek historians, Apollodorius and Diodorus, also alluded to a Pleiadian-Atlantean union.

Other Interstellar Serpents arrived in Atlantis from the star system Sirius, bringing the esoteric tradition of alchemy: the secrets of uniting the polarity. Their symbol was an equilateral triangle with an all-seeing eye in the center. The Sirians began the Great White Brotherhood, which eventually made its way to Egypt, and was later passed on through Freemasonry and the Illuminati.

The Sirians are also well known by the Greeks, the Dogon of Africa, and the Cherokee. The Sirians are mentioned in *The Book of Enoch* as a group of "fallen angels" who came to Earth during an early phase of Earth development. This story later evolved to the story of "Lucifer" and the fallen angels. These are actually the Nephilim, the "People of the Rocketships," as the Sumerians called them. These are the Anunnaki which have been made the most popular group of visiting extraterrestrials by author Zecharia Sitchin.

Another ET Serpent group which arrived on the continent of Mu were described as Cyclopeans. Polynesian legends teach that these gods with golden skin arrived in flying machines and built pyramids throughout the Pacific. It is said that this group has serpent-like bodies and a single psychic eye in the middle of the forehead.

Another ET Serpent group came to Earth from Venus. The androgynous Kumaras were a brotherhood of immortal adepts, or Sons of God. Hindu legends call them Avatars, or Saviors. As Pinkham writes (p. 13):

> The arrival of the Venusian Kumaras on the continent of Mu was pivotal to the unfolding evolution of the human species during the 104,000 year cycle. They brought to Earth the secrets which would assist in uniting the polarity within and lead to the full awakening of unconditional love within the heart of the chakra, the ultimate goal of human existence. The Kumaras were the masters of love and their planet, Venus, is the eternal planet of love.

The Kumaras established mystery schools on Mu and Lemuria, and their Venusian compatriots, the Hathors, also arrived on Earth to serve as teachers and rulers in Lemuria, and later in Egypt. As Pinkham writes, the Lemurian culture was one of the most spiritually advanced cultures on Earth. They were Sun worshippers, and their symbol was a golden sun disc, which represented the transcendental Spirit. The civilizations of Mu, Kumari, and Lemuria were part of the Solar Brotherhood, which later settled in the Peruvian Andes. The Lemurians were said to be spiritually evolved beings, some of whom existed in high frequency physical forms, or "dragon bodies of pure life force."

As Pinkham explains, just before the destruction of Lemuria and Atlantis by an Earth pole shift, some Serpents of Wisdom migrated to pre-determined Earth locations. The Lemurian adepts migrated to the Pacific rim region and founded colonies devoted to spirituality and harmony. The Atlanteans migrated to the Atlantic region, founding empires devoted to intellect and technology.

The Atlanteans established colonies and trading posts in North Africa and the Mediterranean. These early people are known as the Phoenicians, the Basques, and the Tauraks (Tuaregs), the current nomads of the North African Sahara. Other branches of Atlanteans sailed the Mediterranean and settled along the Nile River Valley. According to Pinkham, the Atlantean colonists began to settle in this region between 28,000 to 10,000 years ago, after the Atlantean deluge, but before the Biblical deluge.

Pinkham writes that the Serpent colonists of Mesopotamia, the Anunnaki, were known in both Sumerian, Babylonian, and early Greek texts. The Anunnaki were said to have fish-type bodies with scaly skin. The Greeks called them the "repulsive" or "dragon faced Annedoti." It was said that they arrived at four different intervals, about 30,000 years apart. Pinkham writes, "while some waves of Anunnaki probably came from Sirius, some Sumerian texts seem to imply that at least one delegation came from Atlantis." Atlantis is actually memorialized in Sumerian texts as NI-DUK-KI, a paradise surrounded by water before the deluge. Pinkham also points out that the Sumerian city of Eridu means "home built far away," which may indicate that Eridu was built far from the first home on Earth: Atlantis.

Other groups from Atlantis migrated to the British Isles, and these groups became the Druids and the Celts. According to James Churchwood, in *The Children of Mu*, the migrating serpents of Mu colonized India, China and Tibet, as well as creating a network of caverns and tunnels underneath China and Tibet. Pinkham writes that the Americas, in particular, were a stepping stone for Serpent colonists from the Motherlands of Lemuria and Atlantis. As well, the Americas were "a favorite destination for ET Serpents arriving from Venus and other parts of the cosmos." The name "Amaraka," according to Andean elders, means "Land of the Wise Serpents." The language of the Inkas, Quechuan, is said to be a derivative of the language spoken in Lemuria.

Many Serpents of Wisdom settled in Mesoamerica. The colonizing serpents were the well known leaders, Pacal Votan, Itzamna, and Quetzlcoatl. In addition, many American Indian tribes maintain their origins as the lost Motherlands of Atlantis and Lemuria. In general, Pinkham notes, east coast tribes were settled by Atlanteans and west coast tribes were settled by Lemurians. For instance, the Apaches of Arizona claim to be descended from the Atlanteans who colonized Peru, and the Sioux claim to be descendants of the Turtle branch of the Atlantean Serpents. The Algonquins claim to have come directly from the colony known as "Pan" in Atlantis. The Oklahoma and the Annishnabeg tribes also claim to have come directly from Atlantis. The area of Mount Shasta in Northern California is a notable base of migration of the Lemurians. As the Rosicrucians assert, many Kumara immortals relocated to Mount Shasta just before the deluge that sunk Lemuria, since this mountain range was once attached to the continent of Mu.

The Serpents of Wisdom of Lemuria and Atlantis founded colonies and kingdoms all over the Earth. Until the modern Christian era, which effectively squelched this culture along with all information pertaining to it, these cultures were a worldwide phenomenon. It has been noted that early biblical scholars removed mention of the amphibious and reptilian nature of the founding gods of ancient civilizations.

The Dragon Culture built previously well-known megalithic structures and pyramids which were connected by "an etheric highway of dragon lines." These dragon lines were powerful, naturally-occurring vortexes around the Earth which encircled the planetary electromagnetic grid lines. The pyramidal structures which were strategically built on powerful vortexes, functioned as communications "antennae" between Dragon Communities. This Earth grid was naturally tied to a larger Extraterrestrial Grid of Interstellar and Interplanetary communication lines. Pinkham writes (p. 74):

> These celestial or axiatonal lines emanate out of planets and star systems before eventually arriving on Earth to connect with the planetary grid. Ultimately they unite with the human etheric grid of subtle energy vessels, thereby uniting each Earth human with the larger Galactic Community.

The Galactic Dragon Community, Pinkham explains, aligned these "axiatonal" lines, which streamed to Earth from their home star systems, in order to establish an Interplanetary and Intergalactic communication link. He writes:

> In order to be fully connected to the Extraterrestrial Grid, Serpents of Wisdom around the Earth strategically oriented their pyramids, mounds and temples to significant star positions such as Sirius, Arcturus, the Pleiades, and Orion.

Planet X

In an interview entitled "Planet X and the Return of the Gods" published in *Newspeak* magazine (1995), Sitchin stated that "the conflict between creationists and evolutionists would evaporate once evolutionists recognized the scientific basis of Genesis and creationists realized what its text really says." Sitchin clarified that "cosmic cultures already met about 300,000 years ago." He reiterated the findings of his many years of research into Earth's prehistory, stating that the civilization in Sumer was the eventual result of various genetic manipulations and cultural indoctrinations carried out by a race of extraterrestrial space travelers.

Sitchin believes we are not alone in our own solar system, and that the ancient Sumerians refer to a planet which they called Nibiru. Nibiru is *The Twelfth Planet* within our solar system, counting our Moon and our Sun. The infra-red station Iris sighted a planetary body in December of 1993, and astronomers have named this body Planet X, both because it is the unknown and because it is, in our science, the tenth planet. If this is the planet Nibiru of ancient "myth," we will soon be informed that the reason this planet has not yet been noted by astronomers is that its huge elliptical orbit is on a 3,600 year cycle. This twelfth member of our solar system is discussed in detail in Sitchin's book *The Twelfth Planet*. Sumerian writings indicate that the people who live on this planet regularly visit Earth at intervals when the planet swings physically closer and, likewise, do not visit during those intervals when the planet is taken toward its apogee away from the Sun.

Earth Chronicles, Sitchin's term for the Sumerian writings, explain that the Anunnaki originally came to Earth to obtain gold which they needed for the survival of their planet. He writes in *Wars of Gods and Men*: "at no time during the millennia that followed were these visitors to Earth ever shown wearing golden jewelry." The search for gold was most certainly not a frivolous enterprise; this precious metal may even have been a requirement of the space programs of the Nibiruans, perhaps for their instrumentation and vehicles. But Sitchin has deduced that the needs of the space program were probably secondary to the purpose of "suspending the gold particles in Nibiru's waning atmosphere and to shield it from critical dissipation." Although to date the specific reasons for needing the gold have not been discovered, Sitchin surmises that a planet with such a vast orbit would have the serious problem of heat loss, particularly during its apogee away from the Sun. To retain internal heat, he surmises, you need an atmosphere that acts like a greenhouse effect.

Sumerian accounts explain that the Anunnaki based themselves in Mesopotamia, planning to extract the gold they needed by laboratory methods from the waters of the Persian Gulf. Most of the Earth's gold is in the seas, Sitchin explains, but the problem is how to extract it from billions of tons of water. The Earth Chronicles report that the Nibiruans could not obtain the gold in sufficient quantity this way, and were forced to come up with an alternative method. They knew that gold was abundant in "the Primeval Source" on the continent of Africa, but it would be necessary to drastically alter their approach to obtaining it. The new approach would entail switching from a sophisticated technological "water-treatment process" to "back-breaking toil below the surface of the Earth." The new enterprise also required that more Anunnaki manpower be sent to Earth to settle a mining colony in Africa, as well as to build a fleet of ships to carry the ores to expanded treatment facilities in Mesopotamia. The decision was made to settle in for the long haul.

The arrival of the Anunnaki to Earth for the specific purpose of obtaining gold appears to be premeditated. How did the Nibiruans know that gold and other metals were available on Earth? In his book *Wars of Gods and Men,* Sitchin suggests a few options: They may have probed Earth with unmanned satellites, much like we do now, or they could have sent manned missions to survey Earth in earlier times. In fact, some Sumerian texts mention the Nibiruans landing on Mars as they traveled from Nibiru to Earth. Just as we can analyze the chemical composition of the planets we probe, the Nibiru knew from earlier manned or unmanned ventures where the gold they needed was located on Earth. As Sitchin also reports, a particular Sumerian text refers to an earlier visit to Earth by a deposed ruler of Nibiru. This chronicle, entitled *Kingship in Heaven*, tells of the usurpation of the throne by Anu and of the escape to Earth of Nibiru's deposed ruler. This text records the arrival of a spaceship bearing Nibiruans prior to the mission for gold chronicled 432,000 years before the deluge. It has not been ascertained, but gold may have been discovered at this time. It is important to note that such earlier trips may hold a key to other mysteries as well.

According to the Earth Chronicles, other Anunnaki followed in waves of migration in order to help with the mining operations in Africa. The Babylonian *Epic of Creation* describes the instructions and assignments given to the "rank-and-file gods" involved in the settlement of Earth. The Igigi ("Anunnaki of Heaven") were 300 "true astronauts" who stayed aboard the huge spacecraft which orbited Earth. The smaller "shuttlecraft" shuttled the lesser gods to Earth in groups of 50. The *Epic of Creation* states that the Igigi were never encountered by mankind, but the 600 "Anunnaki of Earth" were the settlers who came to build the cities, work on irrigation and navigation projects, excavate and transport various metals and minerals, and work the gold mines of Africa.

In Mesopotamia, the Sumerians record, five cities were built. Each of these cities had a specific function. Eridu was the original gold-extracting facility by the water, another was the center of metallurgy, another was the Beacon City where the shuttlecraft landed, the fourth was the Bird City or landing place, and the fifth was a medical center. A sixth city was built later expressly for the Igigi, the 300 astronauts who remained in constant orbit above the Earth. Processed ores were shuttled from Earth using the orbiting space platform as an intermediary post, and later transferred to spaceships which carried the product to Nibiru on its near orbit. In addition, a seventh city was built on an artificially raised platform and is described as a sort of "mission control center." These cities were arranged on a ground plan which was detailed in post-deluge chronicles.

A Rock and a Hard Place

Outside of the headquarters for the National Mining Association, located at 1130 17th Street in Washington, DC, a large banner reads: "Everything Begins With Mining." On the banner, the planet Earth is symbolized. The water is blue and the land is gold. I wonder if this organization has any idea of the bizarre truth behind their motto.

According to Sitchin, after working the mines "for forty periods" and "eating dust with their food," the gold miners mutinied. The *Atra-Hasis* chronicle describes the violent rebellion of the Anunnaki miners. They were between a rock and a hard place: for they were being overworked, yet if they abandoned the project they would risk the survival of their planet. Anu, the great lord of the Anunnaki, was called to Earth immediately. The chief scientist among them, whose name was Enki, proposed a solution. He proposed that the Chief Medical Officer, their sister Sud, create a Primitive Worker to labor for them. The *Creation of Man* texts describe in detail the process of creating *The Adam* some 300,000 years ago. Sitchin quotes from the text: "The being already exists. All we have to do is put our mark on it." When you read the text, Sitchin indicates, you realize they are talking about genetic engineering.

In his book, *The Twelfth Planet*, Sitchin divulges that prehistoric mine shafts in southern Africa have produced artifacts dated as old as 50,000 BC. Notched bones, indicating an ability to count, and bedding materials of grass and feathers, indicate that people worked mines in Swaziland between 70,000-80,000 BC. Some ancient gold mines in Swaziland and Zululand have been confirmed to be about 100,000 years old, containing shafts going down fifty feet and evidence of "mining technology." In his book *Indaba My Children*, Credo Vusamazulu Mutwa recounts Zulu legends which hold that the ancient gold mines of Monotapa in Zimbabwe were worked by "artificially produced flesh and blood slaves created by the First People." These Zulu legends also tell of slaves who battled with the Ape Men when "the great war star appeared in the sky."

Sitchin explains that the Bible says: "God said to Adus (plural of Elohim), let us make The Adam (a *specific* being, not mankind) in our image and after our likeness." Sitchin believes that the term "in our image" means *physically* and "after our likeness" means *eternally*. Thus, the Sumerian writings record the story that mankind was created by an advanced civilization of genetic engineers after their image and likeness.

The Cloning of Mankind

According to Sitchin's translations published in *Wars of Gods and Men*, Sumerian texts describe the process whereby the semen of a chosen male Anunnaki (translated as "Those Who From Heaven to Earth Came") was combined with the egg of a female Earth hominid. The fertilized eggs were then implanted into the wombs of female Anunnaki "Birth Goddesses." In this way, the texts explain, the Anunnaki began a slow process of genetically engineering the Lulu Amelu ("Mixed Worker") by binding upon the less evolved beings the "mold of the gods." This feat, however, was not a quick solution, and there was "considerable trial and error to achieve the desired 'perfect model' of the Primitive Worker." It was truly a slow process, for the hybrids were unable to reproduce and the wombs of fourteen Birth Goddesses, seven for female offspring and seven for male offspring, had to continually serve as the containers. Eventually, the workers were put to work in the mines.

The Birth Goddesses could not be expected to continue in their 'labors' indefinitely. The *Creation of Man* texts describe a second wave of genetic manipulations which would enable the beings to procreate on their own, since the initial hybrids were not able to do so. Sitchin points out that cloning requires the least differentiated cells, so the stomach cells are used. Thus, the biblical tale of the creation of Eve from Adam's side is most interesting. In his book *Genesis Revisited*, Sitchin describes the second wave of genetic engineering, which was an operation during which a rib was removed from The Adam—the first man. The Biblical story of the creation tells that man and woman were not created at the same time but, rather, the female was created (or cloned) from the male.

This would seem to make sense in the sequence of events necessary to create a viable species from a hybrid. Sitchin writes: "to be able to have offspring, Adam had to mate with a fully compatible female." He adds: "the tale of the Rib reads almost like a two-sentence summary of a report in a medical journal [describing] a major operation of the kind that makes headlines nowadays when a close relative donates an organ for transplant." The donor in this case was the first-created man, Adam. According to Sitchin's research, Adam was given a general anesthesia in order to remove one of his ribs, and "something was extracted" from the rib in order to "construct" the female.

Sitchin explains, since it is apparent that female workers already existed, something was required from the male in order for the female to become compatible for mating purposes. The clue to what the rib supplied, Sitchin writes in *Genesis Revisited*, lies in the translations of the meanings of the Sumerian words for *life*, *belly* and *clay* (perhaps bone marrow?). Although the original Sumerian texts describing the process have not yet been found, Sitchin surmises that it had something to do with the genetic compatibility of the sperm and the egg. Since pairing takes place when a reproductive cell is fertilized by a sperm cell, the key to reproduction, he explains, lies in the fusion of the two single sets of chromosomes. If their number and genetic code differ the chromosomes cannot combine. Sitchin suggests that perhaps "there was a need to overcome some immunological rejection by the female workers of the males' sperm."

The Anunnaki gave to Adam and Eve and their progeny both intelligence and "knowing": the ability to procreate sexually, but did not pass on their long life span. Since one Earth year is equivalent to 3,600 years on Nibiru (the time it takes for one complete orbit around the Sun), the Anunnaki had incredibly long life spans. Even the ages reported for biblical personages (Methusela's 900-odd years) have been shortened by some sixty percent from the original Sumerian writings, Sitchin reports. In an interesting switch, according to Sumerian writings, another of the Anunnaki creations received the long life span of their creator, but did not receive the ability to procreate.

The Androgynes

In his book *Divine Encounters,* Sitchin reports on the "divine emissaries" of the Anunnaki 'gods'. These divine emissaries can be traced to the gods of Sumer, the extraterrestrial Anunnaki, who "employed emissaries in their dealings with one another rather than with Earthlings." The most well-known vizier of the great gods was named Papsukkal, which translates as "Father/Ancestor of the Emissaries." His diplomatic mission was to convey the great god Anu's advice and decisions to the Anunnaki leaders on Earth. However, there was something particularly unusual, Sitchin points out, about Ninshubur, the emissary of the goddess Ishtar, who sometimes appeared to be masculine and other times is understood to be feminine. Sitchin queries: was Ninshubur asexual, bisexual, androgynous, a eunuch, or something else? Since the physical descriptions of these entities often shifts from tale to tale, Sumerologists are often unable to accurately translate the sex of the emissaries. As an example, the curious asexuality or bisexuality of Ninshubur is "reflected by her/his contesting with other beings" who, Sitchin believes, also "seem to be neither male or female as well as neither divine nor human, a kind of android—automatons in human form."

As Sitchin maintains, to rescue Inanna from the "Lower World," the Anunnaki god Enki created androids who could safely go to this "Land of No Return." Sumerian texts describe how Enki created "two clay androids, and activated them by giving one the Food of Life and the other the Water of Life." As Sitchin relays in *Divine Encounters*, Sumerologists have left untranslated the names of these emissaries, Kurgarru and Kalaturru, since their literal translation is a bizarre reference to inadequate or broken sexual organs: "one whose opening is locked" and "one whose penetrator is sick." Sumerologist Leo Oppenheim has noted that the main attributes which allowed the androids to enter the Lower World was that they were neither male nor female, and were not created in a womb. Oppenheim also discovered a reference to the gods' ability to create robots in the Babylonian creation epic the *Enuma elish*. In this text, Marduk said to his father Ea/Enki, "I shall bring into existence a robot; his name shall be Man ... He shall be charged with the service of the gods." But Ea suggested to "put the mark" of the gods—their genetic imprint—on a being which "already exists."

In the Akkadian version of this story, Ea/Enki created only one being to save Ishtar, and this being was translated as a "eunuch." But, as Sitchin points out, that term, *assinnu*, is literally translated as "penis-vagina;" thus "bisexual" is a more adequate description of this creature. Such androids are also described in Diane Wolkstein's *Inanna, Queen of Heaven and Earth*, as a class of emissaries called Gallu, usually translated as "demons"—which had been created by Enki. The Gallu were described as not being made of flesh and blood. It is written in the Sumerian texts that they were beings who "have no mother, have no father, neither sister or brother, nor wife or child; they know not food, know not water. They flutter in the skies over Earth like wardens." These beings are variously referred to as "angels" and "demons" in the Sumerian literature.

Ancient texts indicate that the Gallu did not resemble the divine Sukkals, the emissaries of the gods, but they did carry the same weapons and used the same "magical" implements. The viziers of the Anunnaki 'gods' were said to carry an instrument which is described as a "wand-weapon," and wore oversized eye goggles which may have protected their eyes from potentially harm-ful rays emitted from the hand-held wand. It was said that from his "ziggurat in Nippur" the great Mesopotamian patron deity Enlil could "raise the beams that search the heart of all the lands," and that he also had "eyes that could scan all the lands," and a "net that could ensnare unauthorized encroachers."

Other gods of Sumer were said to be "armed with the weapon which tears apart and robs the senses" and with a "brilliance that could pulverize moun-tains," as well as "a weapon with 50 killing heads" and a "thunder-stormer which scatters the rocks." Sumerologists have suggested that these patron deities sometimes provided such weapons to Earthly kings, but Sitchin discuss-es the plausibility, in *Divine Encounters*, that use of the weapons or "magical wands" was limited to the trusted emissaries, or viziers, of the Anunnaki gods rather than to Earth humans.

Ongoing Genetic Research

Following this line of inquiry into the possibility of the existence of space travelers, we eventually come to the question: Who are the alien grays and why is it that we do not resemble them? Sitchin indicates that such beings were depicted in antiquity and they were referred to as the "emissaries" of the Anunnaki, employed by them both before they arrived and after they left. He indicates that the term "angel" is derived from the Hebrew meaning "emissary."

If descriptions of the alien grays are to be taken at face value, we might con-sider that the emissaries are appearing in our skies during these times because the so-called twelfth planet Nibiru (or Marduk) is coming closer to Earth on its elliptical course. And where the Anunnaki go, their ambassadors come before them. Are they checking up on our genetic viability? Are they continuing their genetic breeding experiments? Are they looking for fresh stock in order to carry the seeds of a new kind of human to yet another world? Given the warlike nature of Earth humans and the fact that we have discovered the atom bomb, can this new mating project be considered a necessary one for the continuation of our rare species?

The alien gray entity, currently said to be visiting the planet, fits the descrip-tion of the androids or viziers of the Anunnaki. The features of their "skin," described as having an oddly unnatural quality and "fit," along with their gen-derless physique, are the most prominent indications. Also strange is the lack of evidence of aging, breathing, consumption of food or water, and the lack of hair and other mammalian features on an otherwise mammalian-like life form. Also bizarre is a stride which has been described as more like a "glide." Other inter-esting reports are the exhibition of strength without an overt musculature, and the straight fit of the neck into the head and torso, as described by David Jacobs. Also odd is the conveyance of telepathic communication through the eyes, the characteristic non-movement of the eyes and face, and the lack of emotional cues evident in the facial area. In short, the grays come to mind as the beings described by Sumerian scholars, the Gallu, who "have no mother, have no father, neither sister or brother, nor wife or child; they know not food, know not water."

Although they too were created by the Anunnaki, Sitchin is not sure of the biological relationship, if any, between Earth humans and the grays. In ancient writings, such beings are variously described as "angels" and "demons" and are quite distinct from humans. In particular, the Sumerian writings describe the process of the creation of humans, The Adam, by genetic combination of the DNA of the hominids which existed on the planet and the DNA of the Anunnaki, with a second phase of genetic manipulation which gave them "the knowing," or the ability to procreate. Conversely, these texts depict the androids as being "fashioned from clay," or non-living material, and are depicted as having sexual organs that do not work and, thus, they are unable to procreate.

Humanoid Robot Projects

A web site devoted to Sumerian mythology queries why the Anunnaki would not have used robots to work the African gold mines instead of creating a real human work force. This is a good question. It is reported that the Anunnaki initially thought about creating a robot to assist them in mining operations. As discussed earlier, the Babylonian *Enuma elish* quotes Marduk as suggesting the creation of a robot who would be "charged with the service of the gods." Why this idea was outvoted is not clear, but it was decided instead to "put the mark" of the gods on a "being which already exists."

The robot query brings into question just how advanced the technology of the Anunnaki might be in comparison to present Earth civilization. It also brings up the question of which feat would be easier to engineer: a viable "test tube" pregnancy between incompatible species, or a robot mine worker somewhat capable of independent thought, with the ability to perform the physical activities of climbing, walking and utilizing various tools.

In order to put this into perspective, following is an overview of the seven most high-tech humanoid robot projects now under development in the world. Those are Honda Motor Co.'s P2 and P3 walking android being developed in Tokyo, Japan, the Waseda University Humanoid project in Tokyo, the Tokyo University Saika project, the U.K. Shadow Robot Project's Biped Walker, the M.I.T. Cog project under development in Cambridge, Massachusetts, the ProtoAndroid project in development at Faustex Systems Corporation in Houston, Texas, and the Humanoid Robot System under development at the Advanced Robotics Research Center in Korea. A summary of these systems follows. (See pictures at http://www.androidworld.com)

Honda Motor Company's battery-powered P3 android looks like an astronaut in a space suit. The P3 android looks even more like a "Transformer" toy, except that it stands about six feet tall and weighs about 460 pounds. It walks "like a person" and even does stairs. The android was introduced to the public in Tokyo in 1996, after being under development for ten years. The cost of this project has been estimated by outside sources to be at least $5 million, and at least 100 man years; the staff is unknown.

Honda has recently revealed a second android known as P3. This smaller robot is 5'3" and weighs 286 pounds. Honda's P2 was featured, along with the Waseda University Humanoid, on a PBS program called "Robots Rising" in March of 1998. The Waseda Humanoid android is about six feel tall and weighs over 600 pounds. It can just barely walk. The Waseda Humanoid project consists of four groups working in the areas of Vision and Kansei, Design and Brain, Mechanism and Mind, Mind and Communication.

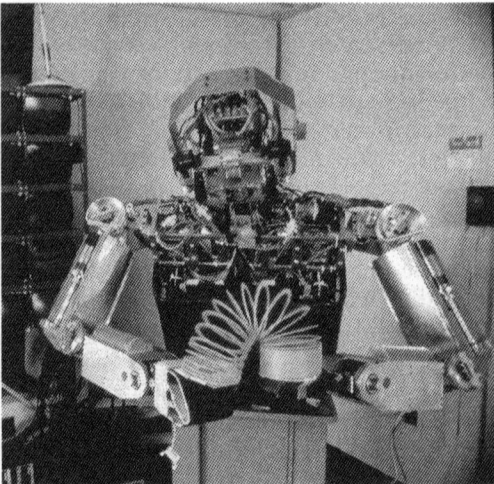

Cog, a humanoid robot being constructed at the MIT Artificial Intelligence Laboratory in Cambridge, MA. Thanks to The Cog Shop at MIT (www.ai.mit.edu.projects.cog/) for permission to picture their creation.

Tokyo University's Saika project is a lighter android of human size. Saika, which means "outstanding intelligence" in Japanese, is a strange looking metal robot consisting of a motorized complex and what might be called a "head," "arms" and "torso" and two great big "eyes." The total weight of its upper body is only eight kilograms. Saika can hit and catch a ball and grasp objects. It is not a biped walker.

The goal of the Humanoid Robot System at the Advanced Robotics Research Center in Korea is to develop a "human-like autonomous robot with sensing and reasoning capabilities." The system under development integrates knowledge of precision mechanics, information technology and artificial intelligence. The goal during the first phase of the project, which began in 1994 and will end in 1999, is to have a "semi-autonomous robot system assisted by tele-operation that can perform dangerous or difficult tasks for humans." The prototype, which is named Centaur, has two arms, four legs and a head, and includes two eyes with robotic vision and a tele-operation system. The four-legged system will contain specially designed feet and posture control design to maintain a stable gait through uneven terrain. Centaur will even walk up and down stairs.

The development expectation of the second phase is to have a two-legged robot with more human qualities, based on the acquired developments of the first phase, by the year 2004. The arms under development for this phase will exhibit 7-degrees of freedom, and will assimilate human arm motions, allowing cooperative tasks between the two arms. A robotic hand with three fingers will be attached to each arm. The fingers will exhibit 4-degrees of freedom, and will have an "artificial skin" to sense pressure, temperature, slippage and tactile information.

The H2-X Hyperkinetic Humanoid Robot under development at Faustex Systems Corporation is touted as the world's most advanced martial arts robot. H2-X is fully operational and has been put to work as a celebrity performance artist in the Austin, Texas area, doing dance and theatre. The H2-X Martial Arts Robot is acclaimed as the "fastest humanoid robot ever." It is faster and tougher than a human, and is safe to bring out into the public arena. It is padded with light weight foams and fabric, and includes other safety features. At 24-degrees of freedom, H2-X is acclaimed as the most advanced of the industrial and academic robots.

Anthropomorphism in Domestic Robots

The Shadow Robot Project has built a prototype bipedal robot called the Shadow Walker. The Shadow Walker is a wooden leg-skeleton powered by air muscles. It stands approximately 160cm tall. The purpose of the research into bipedal walking is to develop the necessary designs and techniques of humanoid balancing and walking. The lower leg is one piece, rather than a two-boned structure, the knee has a simple joint, and the foot has only one toe which is the width of the toe. The hip is a ball-joint which permits three degrees of freedom, and ankle has a double axis design which permits two degrees of freedom. In an on-line document they explain their reason for developing a biped robot. In a section entitled: "Anthropomorphism, or, Why Legs?" they write that the need for anthropomorphism in domestic robots is illustrated by the problem of staircases. While it is not feasible to remove the staircases from houses, it is possible to design robots with stair-climbing attachments, but these are "usually weak spots in the design." Thus, they write, "providing a robot with the same locomotive structures as a human will ensure that it can certainly operate in any environment a human can operate in."

The creators of the Shadow Walker suspect that they will need to add another toe to this design for balancing purposes. They are also working on development of a sensing apparatus for the Walker since the "human balancing process

The Shadow Robot Project has built a prototype bipedal robot called the Shadow Walker. The purpose of the research into bipedal walking is to develop the necessary designs and techniques of humanoid balancing and walking. Special thanks to the Shadow Robot Company Ltd (www.shadow.org.uk).

is greatly aided by the inner ear, which acts as a sensitive 3-axis accelerometer and inclinometer." The robot is also installed with under-foot sensors which are capable of reacting to significant changes, so that "when the robot moves (or is moved) outside the envelope of simply standing, the system will take other actions to restore it to standing up." Thus, the robot is capable of recovering from "small, but significant, perturbations."

However, the Shadow Walker is having a significant problems actually doing what it has been designed to do, which is to walk. The designers describe its attempt at walking after being pushed sideways. The change in mass distribution caused the robot to swing and because there was nothing to stabilize the side-to-side motion, the robot kept swinging over, lifting the other foot from the floor. The robot sort of took two steps, but "luck ran out" and "the third step didn't happen." The robot was caught by its tether and stopped. "However it had taken *two steps*, they write, "Next: to get this to be repeatable."

A UK group called CLAWAR [Climbing and Walking Robots] is a thematic network of people interested in the technology for creating mobile robotics. The commercial applications for climbing and walking robots, they write, were initially recognized by the space industry and nuclear industry in order to address the "identifiable problems of performing maintenance and inspection tasks within hazardous and unstructured environments intrinsically hostile to man." They assert that the nuclear industry maintains a primary interest in the future development of climbing and walking robots.

The Teleman 44 User Group, a European organization, has identified the need for robots which can maneuver in highly dangerous nuclear environments. The success of early prototypes, which include Nero, Lauron II, Robicen, Walking Forest Machine and Robug III, have persuaded the nuclear industry of the potential benefits of this technology. (http://www.uwe.ac.uk/clawar.html)

It is clear from the above descriptions that Earth-based robotics have not achieved the technological level exhibited by the alien "gray" bio-engineered android system. However, the theoretical underpinnings of humanoid robotics serve to illustrate the ways in which a humanoid robot could be of service to its creators. Along these lines, the Cog Shop at the Massachusetts Institute of Technology (MIT) is developing a humanoid robot with four eyes: two with near vision and two with distance vision. Cog has two arms and a torso, but has no legs yet. The emphasis on this humanoid robot is not walking, but thinking. Cog's creators write: "It turns out to be easier to build real robots than to simulate complex interactions with the world, including perception and motor control. Leaving those things out would deprive us of key insights into the nature of human intelligence."

The designers at MIT are mainly interested in exploring human level intelligence issues prior to their simulation of them. The reason for building a human-like robot is also addressed. They write: "The form of our bodies is critical to the representations that we develop and use for both our internal thought and our language." A robot with human-like intelligence must have a human-like body, they assert, "in order to be able to develop similar sorts of representations." Since a robot can have only "a very crude approximation to a human body," they assert, "there is a danger that the essential aspects of the human body will be totally missed;" i.e. that "only the broad outline form is mimicked, but none of the internal essentials are there at all."

In addition, the Cog Shop designers address a second reason for building a human-like robot. The important aspect of being human, they say, is interaction with other humans. They write: "For a human-level intelligent robot to gain experience in interacting with humans it needs a large number of interactions. If the robot has a humanoid form then it will be both easy and natural for humans to interact with it in a human-like way." Thus, if the robot exhibits just a few human-like cues, they assert from observation, "*people naturally fall into the pattern of interacting with it as if it were a human.*" They assert that there would be no reason for people to interact in a human-like way to a "disembodied human intelligence." (emphasis added)

Humanoid Interface Theory

It has been suggested in the UFO literature that the alien gray android may have been designed as a humanoid interface between Earth humans and a non-humanoid race of beings. Therefore, the statement made above with regard to the reason for building humanoid robots, that *people will react to it as if it were human*, is telling. It is clear from the above overview of the current status of robotics that if the alien grays represent the technological level of an advanced extraterrestrial civilization, their Overlords are indeed thousands of years ahead of us. As the creators of Cog assert: "It turns out to be easier to build real robots than to simulate complex interactions with the world, including perception and motor control."

We are now in a better position to speculate on the question of why the Anunnaki chose to create a human work force rather than utilize their robotics technology to mine for gold under the Earth. Apparently, the Anunnaki possessed androids which, if Philip Corso is right, seem to represent "the end process of genetic engineering designed to adapt them to long space voyages within an electromagnetic wave environment." Would such an android be useful in an underworld mining capacity or would it be an undeniable waste of resources?

Even though the androids created by the Anunnaki were apparently capable of taking trips to the Lower World and back, it is not clear that they were real workers. The droids could certainly follow directions, but could the Anunnaki really get a good day's work out of them? Is that what they were built for? Four legged robots could assist in mining operations, just as they do in the nuclear field, but they would still need people to manage them. They may have even had such devices and just didn't leave them laying around for our perusal.

One speculation is that robots would require constant technological maintenance and humans would take care of themselves. In addition, since robots are unable to procreate, the human population would grow as fast as the work required, but to mass produce a fleet of robots would require building materials and computer maintenance systems. It is also clear that the production and use of a robotic army of "smart" genetic collection devices would solve the problem of contamination by infectious microbes, a concern also shared by NASA's Astrobiology motives. In other words, robots wouldn't contract fatal viruses from the planets they visit.

To offset this argument, it is clear that the creation of the human worker was supposed to be a temporary development in order to lessen the toil of the Visitors. Since the "knowing," or procreation, was apparently given to them as an act of dissidence, it would follow that the human workers were designed as an expendable temporary work force. The assignment of the sacred "knowing" to this creation seems to have been a serious moral offense. It would appear that the workers weren't meant to be mass produced, but simply to be incubated slowly in the containers of the Birth Goddesses.

Our inclination to assign morality to an ETI civilization is unwarranted, since it doesn't seem as though the Nibiruans really thought about what they were doing in a moral sense. After all is said and done, it would seem that the creation of human slaves was probably just more feasible under the circumstances, those circumstances being that the Visitors were squatting on an outback planet very far from home, and they didn't have the resources to quickly design and mass produce robots for this particular service. Thus, the decision to perform feats of genetic wonder may have been based on material feasibility and time constraints, as well as the desire to utilize the technology.

According to Sitchin's translations, the initial intention of the genetic breeding program was NOT to populate the Earth, but to orchestrate temporary assistance. The Project simply seems to have gotten out of hand due to the different styles of various managers of the project. After all, they were only human. As the Space Travel Argument suggests, there is always the possibility that a space colonization project can lose control of its "program," or may even be allowed, for unspeculated reasons, to "go into business for itself." Could this be what happened? Indeed, as Neil Freer states, it would appear that the 'gods'of Sumer, the Elohim or Nefilim, "phased off the colony planet," and left "without closing the laboratory door."

As it turns out, the Anunnaki indeed chose to destroy their great civilizations in Mesopotamia with nuclear warheads and take to the skies saying to hell with you all. Physical evidence of nuclear holocaust in Mesopotamia is reported in Sumerian and other historical documents, including the Bible.

Let's Drop The Bomb

As Sitchin relays in his book *The Wars of Gods and Men*, the event which may have started the winding down of the human clock (as discussed in the next chapter) was most definitely *not* a natural calamity. Sitchin recounts Genesis 18 as proof that the destruction of Sodom and Gomorrah was a premeditated act of the Sumerian visitors. In addition, he clarifies, it was "an avoidable event," as opposed to "a calamity caused by irreversible natural forces." And it was also a "postponable" event. Following is a quick overview of how this story is relayed both in the Bible and in Sumerian texts.

In the 24th year when Abraham was 99 years old, he was paid a visit in his encampment near Hebron. The visitor gods were suddenly "stationed upon him." They relayed the message that the Lord was threatening to destroy Sodom and Gomorrah because of the sin and wickedness of the people. Abraham tried to negotiate. He asked, if 50 righteous people could be found, couldn't this event be forestalled? He continued to negotiate until he had the number down to 10. Yet, he was still unable to avert the catastrophe.

As Sitchin explains in *Wars of Gods and Men*, it has never been doubted by orthodox and fundamentalists that "fire and brimstone" poured from the sky to obliterate these two sinful cities. However, natural calamities, such as earthquakes or volcanoes, are typical translations of this event. The event of Genesis 18, however, may be more aptly described as a nuclear holocaust. The evidence that the gods of Sumer dropped The Bomb on humankind has become obvious by studying satellite photos of the Sinai peninsula, Sitchin tells us. He explains, the white limestone mountains surround the plain in an oval shape, yet the black color of the floor of the plain stands in stark contrast. As Sitchin explains, "black is not a natural hue in the Sinai peninsula," with its reds, yellows, browns and whites. The black color is derived from millions and millions of small pieces of blackened rock strewn across the plain.

Sitchin also describes Sumerian "lamentation" texts which vividly describe an atomic explosion: awesome weapons were launched from the skies, spreading rays "toward the four points of the earth, scorching everything like fire." A dense cloud rose up into the sky, followed by a rushing tempest. A deadly cloud was carried eastward toward the Zagros Mountains, affecting all of Sumer from Eridu to Babylon. The lamentation texts also state that a radioactive whirlwind (the "Evil Wind") began to spread west, and Sumer itself was left desolate. As Sitchin asserts, the texts describe more than just destruction; they describe sheer desolation. The cities remained, but the people and the animals died horrible, suffocating deaths, the rivers became "bitter," and the fields "grew only weeds." This was a calamity "unknown to man;" an unseen death that creeps through the streets, passes through walls, and blows in the "howling" wind. The lamentation texts also describe in detail the horror caused by the Evil Wind let loose on Mesopotamia by the gods of Sumer as it spread from one city to the next.

Sanskrit writings also describe unnatural nuclear devastation in ancient times. Occult traditions teach that the legendary lands of Lemuria and Atlantis were destroyed by nuclear weapons. In his 1963 book, *One Hundred Thousand Years of Man's Unknown History*, Robert Charroux wrote that the nature of these cataclysms and the science that spawned them are known to initiates of secret societies, as well as the popes of the Catholic Church. The ownership, origin and history of these ancient weapons of destruction is a tightly controlled religious secret. (see *Cosmic Test Tube*)

Apparently, none of the gods expected this awesome show of desolation and death, and they themselves had to abandon their beloved cities, as described in a Sumerian text called "Lamentation Over the Destruction of Ur." Certain groups of humans who had been forewarned also had a chance to get away, to run for the hills and keep on running.

It is just this "throwback" situation which confounds our efforts to piece together the pre-history of the human race. As we will see shortly, highly advanced people were forced to live as survivalists, in caves and other rock shelters, after their cities were destroyed. And, as Velikovsky has pointed out, each of many cataclysms which hurled humans back to a new "stone age" were considered by those people to be "Time Zero." Therefore, it should be no mystery that certain American Indian legends teach that the First People came down from the sky on a rope and landed in a boat. It should also be no mystery that certain aspects of culture, such as art and myth, seem to be beyond the capacity of their bearers to devise. The Extraterrestrial Hypothesis provides an explanation for these mysteries, and it is all wrapped up in a heavily veiled catastrophic pre-history.

It would appear that, since mankind had aided them in the work of re-establishing civilization on the planet after the great deluge, the Visitors indeed decided to let their project, in the words of The Space Travel Argument, to *go into business for itself*. The result is a civilization which is laboring under the false presumption that it has *evolved* without design or purpose on a lone, outback planet far away from any exogamous intelligence. The next chapter will explore how the gradualistic scheme of Darwinian evolution has helped to turn this trick.

Chapter Seven

The Extraterrestrial Hypothesis and The Timeline of the Emergence of The Human Form

Who came up with Person Man, degraded man, Person Man?

Linnell/Flansburgh,
"They Might Be Giants" CD

In his book *The Twelfth Planet*, Sitchin posits that the Nefilim (Anunnaki) "did not create the mammals or the primates or the hominids, or the genus Homo," but rather, created the first *Homo sapiens*. Sitchin deduces from the Sumerian writings that the Anunnaki arrived about 450,000 years ago, just before the warmer climates of the interglacial period occurring about 435,000 years ago "brought about a proliferation of food and animals," and "speeded up the appearance and spread of an advanced manlike ape, *Homo erectus*." Cylinder seals of the Sumerians depict the "shaggy ape-man among his animal friends," and complain that it let the animals loose from Anunnaki traps. Another Sumerian text states that "the Mother Goddess gave to her creature, Man, 'a skin as the skin of a god,' a smooth, hairless body, quite different from that of the shaggy ape-man."

Sitchin contends in this book that the space travelers took the species *Homo erectus* and "implanted on him their own image and likeness." Importantly, he asserts: "Evolution cannot account for the appearance of *Homo sapiens*, which happened virtually overnight in terms of the millions of years evolution requires, and with no evidence of earlier stages that would indicate a gradual change from *Homo erectus*."

On the other hand, in his book *The Gods of Eden*, William Bramley has queried whether it was the species *Homo sapiens* which was the guinea pig for the genetic engineering of the Anunnaki gods, and whether it was the species *Homo sapiens sapiens* which was the end result of these manipulations. In support of his thesis, he quotes the *Encyclopedia Americana*'s interpretation of the fossil record of *sapiens sapiens* as appearing "with seeming suddenness just over 30,000 years ago, probably earlier in eastern than in western Europe."

Which Shaggy "Ape-Man" Was It?

It is commonly agreed that *Homo sapiens sapiens,* fully modern humans, arrived on the scene rather abruptly only about 30,000 years ago. Whether or not we agree with the currently accepted evolutionary paradigm, let's see how it squares with the Extraterrestrial Hypothesis, and as well, let's explore a few alternative explanations.

It is commonly suggested in anthropological literature that "bipedal hominids" appeared in Africa between three and four million years ago. Fossils of the genus *Australopithecus* are "well represented by many specimens from various places in Africa," displaying a pelvis which would "accompany bipedal locomotion," and generally a cranial capacity which is "small by human standards but large for a small ape," write Lasker and Tyzzer in a college text book entitled *Physical Anthropology*. Crude stone tools also appeared about 2.2 million years ago but it is generally unknown "whether the earliest stone tools were made by *Australopithecus* or by a more advanced contemporary hominid."

It is clear, however, that these populations were very distinct from modern humans. Extreme variation has been noted in this species, which has led some scholars to believe there were at least two species of *Australopithecus*, while others consider these variations to be the result of differences between local populations, sexes, or variations in the gene pool arising over time. Regardless of these differences of opinion, it is widely agreed, write Lasker and Tyzzer, that "at least the early members of the genus were evidently ancestral to our species." They clarify, "evidence from East Africa shows that divergence somehow occurred within the lineage, since robust *Australopithecus* was apparently contemporary with members of the genus Homo about 1.5 million years ago."

On the other hand, the transition between *Homo erectus* and *Homo sapiens* has been a difficult issue for anthropologists to settle. Importantly, although each "species" exhibits its own distinctive traits, "there is a tendency to exaggerate the differences." Ignoring questionable specimens and limiting consideration to fossils found in Java, Peking and East Africa, write Lasker and Tyzzer (p. 352-354), "the range of variation of many features of *Homo erectus* falls within that of modern *Homo sapiens*." In addition, there is a remarkable similarity between *Homo erectus* finds from Lake Turkana, East Africa, dated 1.5 million years before present and *Homo erectus* fossils discovered in Choukoutien, China, dated 400,000 to 500,000 years B.P. (before present), suggesting a "long equilibrium period during which little change occurred."

We need to remind ourselves that these lineages are, at best, educated guesses; that is, *guesses made by people educated in the Darwinian tradition*. For instance, in his book *Mankind Evolving*, Dobhansky suggests that the evolution of *Australopithecine* to *Homo sapiens* occurred in a continuous lineage within a single gene pool. Stephen Stanley disagrees. In his book *The New Evolutionary Timetable,* he outlines his thesis that "a very small number of discrete, long-lived intermediate species may have overlapped each other." Zuckerman has stated that "attempts to place fossils in an evolutionary sequence depend partly on guesswork, and partly on some preconceived conception of the course of hominid evolution."

In short, it's best not to take any of these contrivances very seriously. The species *Homo habilis* (the "Handy Man"), the Leakey family's 1964 missing link, consisted of a lower jaw with teeth, a collarbone, a finger bone, and some small fragments of skull. For the first time, writes Richard Milton in *Shattering the Myths of Darwinism*, "a new human species was to be described on the basis of teeth and fragments alone, and in circumstances where the association of the bones as those of a single individual was conjectural." It has since been suggested that one of the hand bones is actually a piece of vertebra, two of the other bones may belong to a tree-dwelling monkey, and six other bones came from an unspecified non-hominid.

With regard to *Homo habilis*, its hands and feet are also very apelike, calling into question the human-like picture of the Leakey's "Handy Man," as well as other "supposed human ancestors one usually encounters in Time-Life picture books and National Geographic Society television specials." According to Cremo and Thompson in *Hidden History of the Human Race*, some researchers have even concluded that, "there was no justification for 'creating' this species in the first place." It has been suggested that *Homo habilis* was "mistakenly derived from a mixture of skeletal elements belonging to *Australopithecus* and *Homo erectus*." Others believe *Homo habilis* bones are completely *australop - ithecine*. It is no wonder that Richard Leakey's assessment of the material in this book, displayed on the back jacket, states: "Your book is pure humbug and does not deserve to be taken seriously by anyone but a fool." Could somebody be feeling foolish?

A Gorilla Picnic

In 1994, two arm bones, an ulna and a humerus, were established as belonging to the mysterious creature, *Australopithecus afarensis*. It is typical that an entire evolutionary sequence can be extrapolated from a few bones, sometimes even scattered across a few miles. Such illicit extrapolations can eventually come to describe a creature which "climbed in the trees but also walked on two legs when on the ground." Hmmm. Could these researchers be looking for a "transitional" find?

The best known fossil of *A. afarensis* is the one called Lucy, which you can see at the Museum of Natural History in New York and London. From her glass case, writes Richard Milton, "Lucy peers with an intelligent gaze at visitors, her posture fully erect and humanlike, her hands and feet also short and humanlike." Incredibly, Lucy's apelike appearance was ignored when she was restored to ostensibly lifelike appearance for both of these museum exhibits. In actuality, the hands and feet of this species are long and curved like that of a tree-dwelling ape. The finger and toe bones of this species are "highly curved even when compared to those of a modern ape like a chimpanzee." Milton quips: "Just why Lucy should have been restored to have humanlike hands and feet, contrary to the known anatomical facts, remains a mystery which only her restorers can explain."

Specifically, the rib cage of *A. afarensis* is conical in shape, not barrel-shaped like a human rib cage. Lucy's shoulders, trunk, and waist have a "strong apelike aspect to them." At an international conference held in Paris in 1989, anthropologist Peter Schmid stated that *A. afarensis* "would not have been able to lift its thorax for the kind of deep breathing that we do when we run." He explained that the abdomen was potbellied and there was no waist, thus restricting the flexibility required to accomplish the feat of running. In addition, Leslie Aiello's work on body weight and stature of *A. afarensis* clearly identified it as an ape. The australopithecines were more apelike in their body build; heavily built for their stature, like that of a present day ape. Richard Leakey sums it up: "Australopithecines had been bipeds, but were restricted in their agility; while species of *Homo* were athletes." According to Leakey in *Origin of Humankind*, the inner ear structure of *A. afarensis* has been shown to exhibit semicircular canals which resemble those of apes. The structure of the pelvis and lower limbs also suggest an apelike gait. Leakey also states that *A. afarensis* was *not* a toolmaker.

With regard to different *Australopithecus* finds in South Africa called *P. robustus*, Ian Tattersall—Curator of the Department of Anthropology at the American Museum of Natural History—describes in detail the dentition, skull size and shape and facial architecture of fossils attributed to this 1.7 million year old species. His illustrated compendium entitled *The Last Neanderthal* shows an upper skull of this australopithecene, which exhibits, in his words, a "sagittal crest reminiscent of those in some gorillas." Describing the brain case, Tattersall admits it is "hard to estimate it as a proportion of body size, because not very much of the body skeleton is known."

However, neither the author nor his publisher, Simon &

Darwinian Artistic License: Human-like limbs are falsely attributed to "Lucy," an ape-like *Australopithecus Afarensis*

Schuster, are reticent to reproduce alongside these descriptions an artistic rendition of these supposed "pre-humans," with arms and legs of modern human proportion. A color illustration shows a rather tall and lanky, fully upright gorilla-human couple carrying a gorilla-human child, walking through an open pasture with a fully upright gait. The pastoral scene is complete with frolicking gorilla-human children and several deer: essentially a gorilla-human picnic without the basket, circa 1.7 million B.P. Alongside this scene, Tattersall suggests: "it is likely, if not entirely certain, that these hominids used bones and horn cores for digging." As noted earlier, Leakey has stated that this hominid was "not a toolmaker." Tattersall hasn't lied; he has merely suggested that the hominids picked something up and used it to dig. Chimpanzees do that. But, as sure as they are standing, there *should* be no question that fully upright bipeds would be toolmakers. It goes with the territory. Everything is wrong with this picture.

Such artistic license has been the hallmark of the Darwinian evolutionary paradigm. As Richard Milton explains in *Shattering the Myths*, alongside prints of the busts of Piltdown man, Java man and Neanderthal man, "Darwinian restorations based on fragmentary finds of bones and teeth always manage to convey a distinct 'missing link' quality to their former owners." While the Piltdown man was a hoax in which someone actually put a human skull with the jaw of an ape, essentially the same artistic motif blends humans and apes in popular books and even in museum displays, to give the missing link-look to genuine fossils. The missing link look also includes the illicit addition of human proportions to the limbs, and an upright bipedal gait. Milton adds: "this modern confidence and apparent precision in reconstruction is not based on further discoveries of fact, but takes place *despite* the discoveries of recent decades—that the evidence for humankind's own evolution is actually nonexistent."

When I have had occasion to point out such Frankensteinian artistic license, people have looked at *me* incredulously, and have asked, "you mean somebody makes these up?" Yes, Virginia, they do. It is even more horrendous that museums are guilty of this crime against truth. They should be held accountable for purveying such outright lies to the public under the ostensible purpose of education.

In fact, according to Milton, the status of *Australopithecus* as an extinct ape was actually established as long ago as 1954 by zoologist Solly Zuckerman, who deduced, by measuring the skulls and teeth of a large number of modern apes, human specimens and *Australopithecine* fossils, that the head of this species was balanced like that of an ape, its brain was the same size as a modern gorilla, and its jaw and teeth are predominantly apelike. The same conclusion was reached by Charles Oxnard of the University of Western Australia in 1984. In his book, *The Order of Man*, he deduced that "*Australopithecus* is an extinct ape and is unconnected with humankind's ancestry."

Also, in *Hidden History of the Human Race*, it is maintained that generations of experts have been "wildly mistaken" about the australopithecines. The authors point out that even Louis B. Leakey had concluded that australopithecines were a side branch and not in the direct lineage of *sapiens*. Since *Homo erectus* was thought to be a descendant of *Australopithecus*, Leakey also removed *erectus* from the line of human ancestry. Zuckerman has been almost a lone voice in the wilderness, challenging the assumptions about the *Australopithecus-Homo sapiens* relationship, but apparently this small band of voices has not been enough to challenge the status quo. As he asserts, the voice of higher authority, in due course, has become universally incorporated into all anthropology text books. This voice of authority has managed to keep the human-like view of *Australopithecus* intact in the mind of the general populace.

Incidentally, Cremo and Thompson point to institutions such as the Carnegie Institute and the Rockefeller Foundation as two of the primary sources financing evolution, the Big Bang theory, and the materialistic cosmology in general.

While the current paradigm invites australopithecines to mankind's family picnic, it is clear that defining *Australopithecus* as simply a "biped" is not enough to put this creature in the human family, since there are more aspects of humanness than simply exhibiting some form of bipedalism. In *The Origin of Humankind,* Leakey concludes that the shape of the *Homo* lineage earlier than two million years ago must be regarded as an unresolved question. This is because it remains heavily disputed whether the size ranges of early fossils indicate variation between *males and females* or whether the size ranges indicate *different species*. Under the most ludicrous diagram consisting of two versions of evolutionary boxes with the names: *A. afarensis, A. africanus, H. habilis, A. robustus, A. bosei, H. erectus*, and so forth, Leakey writes: "Family trees. The existing fossil evidence is interpreted differently by different scholars, although the overall shape of the inferred evolutionary history is similar."

ERA	my	PERIOD	EPOCH	
CENOZOIC	2	QUATERNARY	HOLOCENE	
			PLEISTOCENE	
		TERTIARY	PLIOCENE	
	65		MIOCENE	NEOGENE
		CRETACEOUS	OLIGOCENE	
	140		EOCENE	PALEOGENE
			PALEOCENE	
MESOZOIC		JURASSIC		
	210			
	250	TRIASSIC		
	280	PERMIAN	PENNSYLVANIAN	
	320	CARBONIFEROUS	MISSISSIPPIAN	
PALEOZOIC	360			
		DEVONIAN		
	400	SILURIAN		
	440			
	500	ORDOVICIAN		
		CAMBRIAN		
	570			
		PRECAMBRIAN		

The generally accepted Geological Time Scale as found on Dr. Bob's Geologic Time Page (http://oldsci.edu/geology/jorstad/geoltime.html).

Yes, the overall shape *would be similar*, since all of the box designers are working with the same Darwinian, materialist, causal, paradigm! How could we expect the shape of an evolutionary diagram to be anything but a linear, historical and connected movement of the representational animal figures of a powerful Western totem? Interestingly, it's even called a "family tree"!

Forms of Temporal Tinkering

It is questionable, writes Richard Milton in *Shattering the Myths of Darwinism*, that "the geological column is a record of processes taking millennia to unfold; and whether the fossils it contains are a living succession." The various sedimentary rock strata which are piled on top of each other in supposed chronological sequence represent the successive phases of the deposition of sediment. These strata are extensively classified and correlated all over world. Interpretation of this stratigraphy is complicated by the fact that some of the beds have been eroded over and over again, which provides gaps in the sequence. In addition, the Earth's crust has been distorted by folding and volcanic activity. Nowhere in the world, writes Richard Milton, "is there known to be a complete sequence of sediments from the oldest to the most recent."

In addition, most dating scientists practice "intellectual phase locking": the practice of correcting experimental errors on the side of the currently accepted values. When various dating methods produce discordant dates for the same sample, Milton asserts, "the figures are adjusted until they seem right." The most common way to harmonize discordant dates, he explains, is to label the unexpected or unwanted dates as "anomalous." This is why many dating results seem to support each other: the dating scientists have discarded the unwanted, anomalous results. Dates must land inside a certain "ballpark" and must corroborate other established dates. A scientist who obtains a date which is way outside the ballpark would not rush to publish such a finding, while a date which coincides with other findings is published immediately.

By assuming the fact of evolution, scientists can date their hominid finds by morphology, and construct from this a sequential contrivance called evolution. On the sole basis of their commitment to evolution, Cremo and Thompson assert in *Hidden History of the Human Race*, dating scientists decide that a more ape-like specimen should be moved to the early part of its possible date range so that it does not overlap with a more human-like specimen. Likewise, a human-like specimen is moved to a later, or earlier date, within its own possible date range. Thus, the two specimens are separated in time. This orchestration is based on the morphology of the specimens, or "morphological dating." As Cremo and Thompson maintain, "it would look bad to have two forms, one generally considered ancestral to the other, existing contemporaneously." This is the way in which a "temporal evolutionary sequence" is born.

By assuming as factual the evolution of the great ape lineage into the human lineage, the search for the ultimate missing link causes scientists to perform what we might call "temporal tinkering." A tautology is then utilized to enforce this rule: the morphology of fossils is used to select the desirable dates within the possible date ranges of the sites, thus preserving an evolutionary progression in the clay of their minds. This "artificially constructed sequence," which is designed to fit the desired evolutionary paradigm, is then "cited as proof of the evolutionary hypothesis."

By formulating a temporal sequence which presupposes that the hominids *did not co-exist*, this methodology assures that no fossil evidence will "fall outside the realm of evolutionary expectations." This "co-existence" factor is an unhallowed and unacceptable conclusion, and is to be avoided at all costs. Yet, fossil evidence in China, for instance, does indicate that several different types of hominids did co-exist in the middle Middle Pleistocene. In fact, it would appear that humans have been around much longer than we have suspected and, indeed, it is highly suspect that what we are looking at in the fossil record is the co-existence of two separate animals: ape and man.

Dating Homo Erectus?

While supposed evidence of a close relationship between arboreal apes and humans is forced on the public, other evidence for a much more ancient and separate origin of humans remains effectively buried. Cremo and Thompson's findings in *Hidden History of the Human Race* indicate that human origins actually have nothing to do with the great ape lineage.

Evidence hidden away from public scrutiny indicates that humans are not related at all to the great apes, but are a separate species of exceptionally ancient origin. The authors outline the evidence discovered over the years which indicates that humans have a much more remote history than the accepted paradigm would convey. Possible human skeletons from the Eocene and Miocene periods have been discovered. However, the authors write, because these finds don't fit into the accepted scheme of things, they are undocumented, uninvestigated, and conveniently "forgotten." In contrast, they write, "finds which conform to accepted theories are thoroughly investigated, extensively reported, and safely enshrined in museums."

Java Man is an example of the problems involved with the search for humanity's enigmatic "missing link." Modern researchers have pronounced that the so-called Trinil femurs (leg bones) of "Java Man," found in Indonesia, are not Homo erectus, but are the bones of fully modern sapiens. According to the authors of Hidden History of the Human Race, it now appears that "we can accept them as evidence for anatomically modern humans existing 800,000 years ago." Yet, visitors to museums around the world are still treated to casts of these items in the context of Homo erectus, and in the context of fossil evidence for human evolution out of Africa in the established sequence. The twist is, the Trinil femurs are fully modern, but the skullcap attributed to this creature is more archaic. It has been suggested by many that the skull does not belong with the femurs but, rather, with the skull of Pithecanthropus, an extinct ape.

According to Cremo and Thompson, the formation where Java Man was found has a potassium-argon date of 800,000 B.P., but other beds in this formation are dated at over a million years. Fluorine content test results on the bones are consistent with the date of 800,000 years for modern humans in Java. The skull and femurs indicate the presence of two kinds of hominids in Java during the early Middle Pleistocene, one with an ape-like head and the other with legs of modern human type. Therefore, "Java Man" is a totally contrived creature touted as mankind's missing link: *Homo erectus*.

Although Lasker and Tyzzer write in *Physical Anthropology* that "the degree of difference between *Homo sapiens* and *Homo erectus* is seen to separate them as two species of the same genus rather than as two genera," it should be noted that any species designation is an arbitrary one. As also noted earlier, the range of variation of many features of *Homo erectus* falls completely within that of modern *Homo sapiens.* Additionally, there is a problem with dating these fossils since, they state, "all fossils relating to the transition from *Homo erectus* to *Homo sapiens* are too ancient to be radio-carbon dated but too recent to be dated by the potassium argon method."

Nonetheless, the mainstream human evolutionary paradigm calls for a *swift divergence* occurring approximately two million years ago in the hominid line, which, as will soon be illustrated, may represent two entirely different animals. This was ostensibly followed by a long period of *relative stability* in the genus *Homo* until about 400,000 years before present, when a new divergence seems to be recorded. Does "something" seem to have happened at about the time frame which Sitchin suggests for the arrival of the Visitors? Let's explore.

The basic anatomical pattern of *Homo erectus* exhibits a brain of about 900 cubic centimeters, a long and low cranium, a small forehead, a thick skull, a protruding jaw and prominent eye ridges. This pattern persisted, writes Richard Leakey, until about a half a million years ago, followed by "an expansion of the brain during this time to more than 1100 cubic centimeters." Also by this time, Leakey states, "*Homo erectus* populations had spread out from Africa and were occupying large regions of Asia and Europe." He goes on to qualify that *no unequivocally identified Homo erectus fossils have been found in Europe*, but it is *surmised* that they were there because of the technology normally associated with their existence. It should also be noted that the tool technologies which were utilized over a period of thousands of years essentially overlap and are commonly thought to have been borrowed by different types of human variants, including Neanderthals.

Did Homo Erectus Evolve Twice?

"The advancement of knowledge," asserts Van Flandern in *The Anomalist*, "should be the only objective of scientific observation and experimentation, rather than the propagation of commonly-held belief systems." In that regard, the prehistory of *Homo erectus* is now flying in the face of the mainstream paradigm which states that humans came out of Africa. According to new *Homo erectus* fossil finds on the Indonesian island of Java, the origin of mankind has been pushed back to 1.8 million years, suggesting that "either erectus *migrated to Asia* as soon as he appeared in Africa — which is rather unlikely — or that different variants of the genus Homo *evolved independently in different places on the globe."*

Archeologist Yuri Mochanov has spent the last decade excavating Diring, a site along the Lena River in Siberia. His discoveries have forced members of the profession to reconsider the commonly-held paradigm that humankind evolved and dispersed in the warmer climate of Africa, where surely the environment would have been more conducive to survival than under the extreme conditions of Arctic Siberia. This Earliest Paleolithic layer has produced more than 4,000 artifacts, with some 500 of them identified as stone tools.

According to this interview entitled "On Human Origins: Out of Siberia?," published in *The Anomalist 2*, the finding that Stone Age hominids lived in the far north as long as 3 million years ago upsets the mainstream evolutionary paradigm in several ways. While some archeologists are not convinced that the site is 3 million years old, all are convinced that the site is at least 500,000 years old. Mochanov compares the Diring complex to the stone complexes found in Olduvai Gorge in Africa dating from 1.7 to 2.7 million years ago. Yet, until now, he points out, Siberia has not produced stone tools more than 35,000 years old.

This type of stone tool culture has no comparison in Siberia, Eurasia, America or Australia. Diring is one of the oldest Paleolithic sites in the world, dating from between 1.8 to 3.7 million years B.P. It has been established that permafrost conditions did exist in the area at the time, and experts have determined that the average annual temperature in Yakutsk, Siberia was even lower than the present temperature of ten degrees below zero (Celsius). The extremely frigid climate presents "a major stumbling block for the existence of early man in this area." Mochanov suggests it was "the extreme environmental stress of the region that actually *gave pre-humans the fateful genetic nudge* to develop the large brain that defines the genus *Homo*." Therefore, he suggests, mankind may have evolved twice!

This thinking suggests that these university-trained apologists for Darwin are not just encouraged but *trained* to come up with such ridiculous explanations which give evolution a leading role in the grand performance called the "human race." As Phillip Johnson points out, neo-Darwinists have "evolved an array of subsidiary concepts capable of furnishing a plausible explanation for just about any conceivable eventuality;" and these anthr-apologists do not seem to notice that their knee-jerk reactions are problematical to the accidental nature of Darwinian theory: that evolution is *ipso facto* a chain of contingent events which easily could have been otherwise. The concept of survival of the fittest is chaos theory at its finest; it's an absurd dice game.

How could both the desert climate of Africa and the frigid climate of Siberia, two remarkably dissimilar environments, independently and accidentally produce the same rare evolutionary novelty called *Homo sapiens*? Don't forget that this is a novelty so rare that the supposed incremental steps leading to its development could not possibly be repeated given the billions and billions of stars in the Universe! When presented with archeological evidence that humankind may have "evolved" simultaneously in vastly different environments on the globe, a development that goes against evolution's most basic theoretical premise, why is it that the "fact" of evolution itself does not come under scrutiny? Or, why doesn't the correctness of the sacred geological column come under scrutiny? Instead when the pieces do not fit, they are chiseled and drilled and made to fit in the most ridiculous jury-rigged manner.

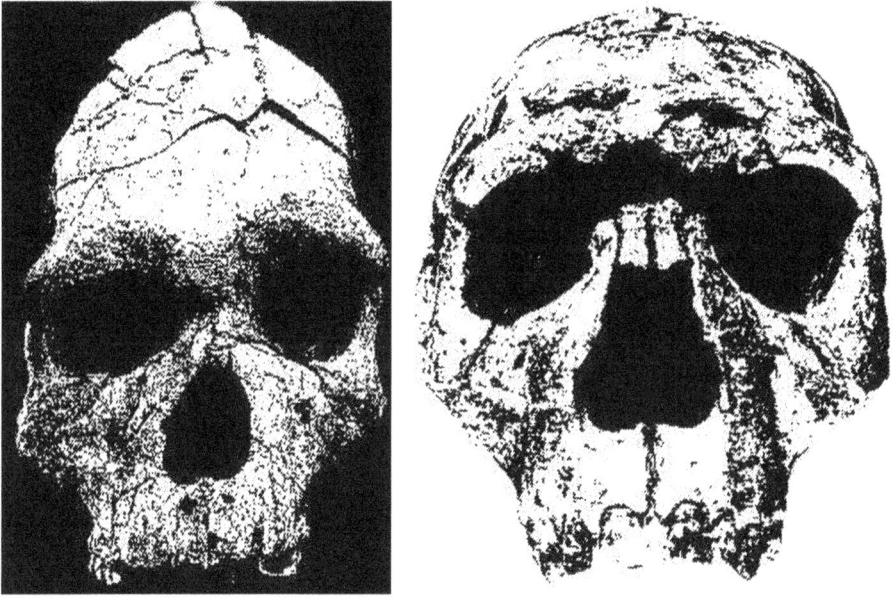

A comparison of the features of the Arago *Homo sapiens* (left) dated at half million years and the Turkana (African) *Homo erectus* (right) dated at one and a half million years. Not shown to scale.

Anthr-apologists tell us on one hand that evolutionary man is the winner of a preposterous survival lottery which, given the incredible odds, should not have even occurred once. Then, when the archeological record doesn't mesh with the theory, the straight-faced suggestion is that the independent evolution of mankind must have occurred twice! There is never the suggestion that the emergence of the human form did not occur without genetic tinkering from the outside, or that it is guided by a currently misunderstood operant force or intelligent factor, or that it is a gift of God, even in the face of evidence that either of these alternatives may be the case. In order to be internally consistent with its current scientific paradigm, the theory must explain a strictly 'natural' cause inherent *within* the operating system: therefore, humankind *accidentally* hit the bull's eye *twice*!

Utilizing the absurdly tautological method of determining the age of rocks by the fossils in them, and the age of fossils by the rocks they're sitting in, existing archeological theory is incompatible with the presence of Oldowan-style tool-makers in the far north 2-3 million years ago. Mochanov asserts, "all skeptics have the same viewpoint; it is impossible because it cannot be so old. But when those skeptics are asked to point at the mistakes we have made, they can't." They can't because they are locked into the tautology of the geological column, as well as to the currently accepted evolutionary paradigm which asserts mankind's linear evolution is traceable from one branch within a specific time frame. Any data that falls out of the limits of this box upsets the entire scheme. Thus, they have painted themselves into a corner.

The Diring pebble tool site presents a problem for the peopling of the Americas and poses questions with regard to the origin of the American Indians. Archeologists generally agree that the American continent was peopled only 12,000 to 40,000 years ago at most. But if people lived in Siberia 2 to 3 million years ago, what prevented them from walking across the Bering Strait land bridge hundreds of thousands of years earlier? Perhaps nothing prevented them.

On the contrary, as discussed by Cremo and Thompson in *Hidden History of the Human Race*, sophisticated stone tools discovered near Mexico City, which rival the best work of Cro-Magnon man, have been given a "lunatic fringe" date of 250,000 years. Also among the discoveries outlined by Cremo and Thompson is an expertly worked "Folsom" blade, discovered embedded in travertine crust from Sandia Cave in New Mexico, which is also said to be 250,000 years old.

The Australian "Home Erectus" Controversy

A 1998 on-line article by Jim Vanhollebeke entitled *Kow Swamp: Is It Homo erectus?* ("A Refutation of the Supposed Insignificance of Certain Australian Hominid Fossils") presents arguments countering the assertion that the well-known but conveniently ignored Australian Kow Swamp fossils, discovered as long ago as the 1880's, are representative of *Homo erectus*. The author argues convincingly that the KS fossils, including Talgai, Cohuna, Nacurrie, Coobool Creek, Kow Swamp, Willandra Lakes, and others, are not *Homo erectus*, but are modern in age, dated about 10,000 to 30,000 years before present.

The KS fossils display primitive or "archaic" features, but are of a very recent age. In addition, other much older human fossils discovered in Australia display a much less "archaic" nature. Therefore, as Vanhollebeke notes, the KS fossils remain as "odd footnotes in the world of Paleoanthropology." They are essentially ignored by most anthropologists. The author writes:

> Accepting these fossils for what they are has been a problem for many anthropologists. Part of this problem, possibly, is the fact that the present aboriginal population in this area of the globe, to varying lesser degrees has been known to exhibit some or all of the traits that make the Kow Swamp type so controversial. This would indicate an obvious line (or lines) of descent. This is not really surprising when the age (or lack thereof) of the fossils themselves is taken into consideration. Obviously the specimens now preserved do not represent the very last of their kind.

Vanhollebeke suggests that direct descendants of the Kow Swamp people would have continued in this isolated region for thousands of years. In addition, it is plausible that the KS populations dwindled slowly, and were diluted by gene flow with other types. Yet the important point to keep in mind is that the

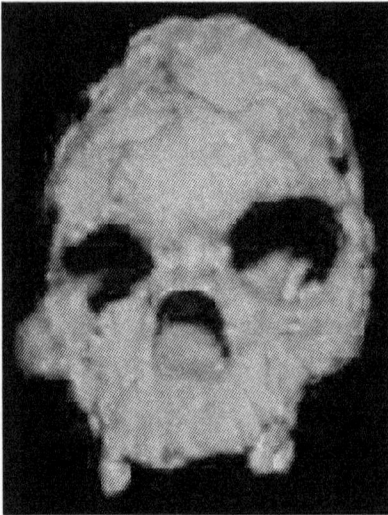

The Australian Kow Swamp fossil dated at 12,000 years shows many similarities the Arago and Turkana skulls on page 171.

KS fossils closely resemble certain living groups of native Australians. Comparing KS fossil skulls to aborigines in northern Queensland [see photos], he writes: "The KS-type fossils are so recent that their unique archaic traits continue to show in living descendants." This can be a "delicate matter," Vanhollebeke writes, in terms of race. However, he admonishes, this is not really a matter of race, since it has long been held that the only way to describe the physical features of certain of Australia's aboriginal populations is "archaic Caucasoid."

If the features of the KS fossils represented Neanderthal characteristics, it wouldn't be such a problem, writes Vanhollebeke. Problematically, the KS fossils more closely resemble *Homo erectus*. As he explains, even late *Homo erectus* has been considered to be extinct for hundreds of thousands of years. But here we have living human beings who resemble these archaic forms.

No wonder it is best to ignore this situation. In addition, Vanhollebeke points out, the recent discovery of "Solo Man," a large brained late *Homo erectus* population living in Java, may actually have survived as recently as 27,000 years ago. Solo Man was previously thought to have been extinct 200,000 years ago. This suggests two things: (1) This fossil represents *Homo sapiens* and not *Homo erectus*, or (2) late *Homo erectus* co-existed with modern man in Southeast Asia. Vanhollebeke asks: "Where are you National Geographic?"

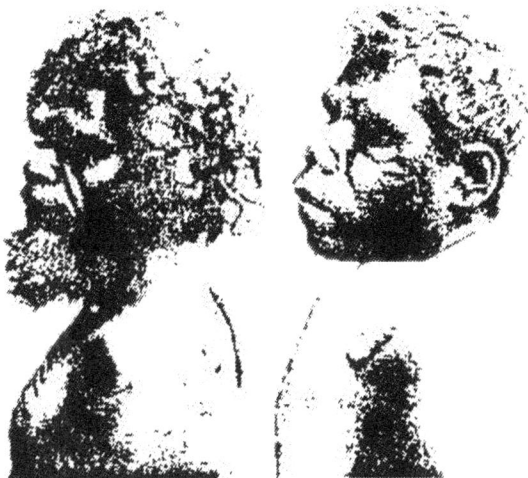

In Thomas Huxley's evolutionary scheme, the Australian aborigines are the "lowest" race. On the contrary, there is evidence that this human type resembles ancient *Homo erectus*, suggesting that *Homo sapiens* is extremely ancient and that there is no such thing as *Homo erectus*.

The fossil record in Australia shows that there were in fact two distinct human populations in Australia during the late Pleistocene (approx. 500,000 B.P.) There was an older yet more modern and gracile type, and there was a much more recent but primitive and robust type. Vanhollebeke writes: "When the media (and world) can marvel at the pile of fragments they call "Lucy," there should be a little awe available for this aberrant populace that never quite went extinct." The story of mankind, he writes, is "sketchy and full of speculation." We must remain open to new ideas when it comes to reconciling these anomalies.

Problems With "Transitional" Types

In their book, *Hidden History of the Human Race,* Cremo and Thompson discuss various skeletal remains of anatomically modern humans, as well as various human artifacts, which have been given dates in the range of 2 to 55 million years, or more. Such discoveries are part of the "hidden history of the human race," and are considered "anomalous" within the context of the currently accepted paradigm of human evolution. As Cremo and Thompson write, in the past, "anomalous" evidence was often "the center of serious, longstanding controversy within the very heart of elite scientific circles." Furthermore, evidence of this kind, the authors assert, is not always of a "marginal crackpot nature." Nonetheless, it is now a matter of course to reject anomalous findings outright, and, further, to "forget" they even exist. Consider, for instance, some of the following evidence hidden in humanity's closet:

- Simple Eolithic implements have been discovered on the American continent in Pliocene strata dated at 2-4 million years.
- Primitive bone implements, as well as shark teeth with holes for use as jewelry and carving implements, have been found in Suffolk, England, in formations dated at 2-55 million years.
- A wooden tool which has been sawn and burned on one end was found in England and dated at about a half million years. It appears that only a metal saw could have accomplished this type of clear cutting, and only *Homo sapiens* could have effected this feat (not *Homo erectus*!)
- Simple chopping tools have been discovered in Pakistan in formations dated at 2 million years. Tools of the same age have also been found in Siberia and

India. (The authors point out that modern tribal people continue to manufacture very primitive types of stone tools.)

• A highly anomalous find of a possible *Homo erectus* fossil skullcap in Brazil challenges the theory that only anatomically modern humans made it to the American continent.

• A modern-type human jaw thoroughly infiltrated with iron oxide was discovered in a quarry at Foxhall, England, in a 16-foot level dated at 2.5 million years.

• A fully modern human skull found in Buenos Aires, Argentina has been dated at least a million years old.

• An anatomically modern human skeleton in natural connection was found on the Italian Riviera in a layer dated 3-4 million years.

• An "atlas," the upper bone of the spinal column, was discovered in Monte Hermoso, Argentina in a layer dated 3-5 million years. Flint tools and intentional use of fire in this area, at the same level, indicates the presence of humans in the Americas at least 3 million years ago.

• A modern human skull was discovered in the Sierra Nevada mountains, under volcanic ash in a gold-bearing gravel bed ranging in age from 9 to 55 million years old. Additional human skeletal remains, and a large number of stone implements have also been discovered in the same beds, so this is not an isolated discovery.

• The California gold country has been a hot bed of human skeletal remains and implements having date ranges of 9 to 55 million years. Stone artifacts and a modern type human jaw were discovered beneath the lava caps in these gold-bearing gravels. In addition, a human leg bone found in these gravels is dated at 8.7 million years.

• In addition, there is strong evidence for the presence of rounded bola (sling stone) makers in Argentina approximately 3 million years ago. The bolas of Miramar point to the existence of human beings of a high level culture during the Pliocene, or even earlier, in South America. Because of their technological sophistication, sling stones and bola stones represent the presence of *Homo sapiens*. Sling stones have been discovered in various places around the globe, including England, East Africa (Tanzania), and Argentina. Stone bolas were used for hunting by wrapping them in leather bags attached to a long cord, and swinging them overhead and letting go. The use of stone bolas necessitates the presence of a leather working culture. The dates of 1.7 to 2 million years are considered "anomalous," since, it is believed, the australopithecines of this age were not toolmakers, and furthermore, they were still confined to Africa.

According to the currently prevailing paradigm of human evolution, the creature called *Homo habilis*, who was not even a tool maker, should have been confined to Africa during this time period. The "standard view," Cremo and Thompson point out in *Hidden History*, is that *Homo erectus* was the first representative of the *Homo* line to emigrate from Africa no more than a million years ago. Anything earlier than this date is considered "anomalous," (i.e. doesn't fit). An indication of the emotional import of such finds in scientific circles is the typical demand for "higher levels of proof for anomalous finds than for evidence that fits within the established ideas about human evolution." Yet, there is no difference in the workmanship of Eolithic implements found in Olduvai Gorge in East Africa and those found in England. If the stone tools of England are rejected as being nature-made, Cremo and Thompson assert, then those of Africa need to be thrown out as well!

The authors of *Hidden History* suggest the incomprehensible: perhaps there were creatures of fully modern type already at Olduvai Gorge in Africa during the earliest Pleistocene era. The reason we can't get around this is because no human fossils are accepted to be that old. But, the authors point out, several human-like femurs discovered in East Africa which were attributed to *Homo habilis* may have actually belonged to anatomically modern humans.

The creature touted as *Homo habilis* has since been shown to have a more ape-like anatomy. Further, Louis Leakey found a bone tool in the same level as the stone bolas discovered in East Africa, which, he stated in 1960, was some sort of "lissoir for working leather." Leakey believed this find suggested a more evolved way of life for the Oldowan culture. Leakey reportedly discovered a fully human jaw at Kanam, East Africa, dated approximately 2 million years.

As we have seen, distinctly anomalous discoveries are not confined to 19th century "nutcases." They have continued, Cremo and Thompson insist, "with astonishing regularity" to the present, but are not recognized for what they are. This is due to the fact that the idea of the progressive evolution of humans from the great ape family guides the acceptance and rejection of the evidence for evolution, and contradictory evidence is pushed to the "lunatic fringe." In fact, evidence suggests that fully modern *Homo sapiens* appears to have existed in Africa alone from very early times, *at least* two million years before present. Taking into account only the more conventionally accepted evidence, Cremo and Thompson have concluded that "the total evidence, including fossil bones and artifacts, is most consistent with the view that anatomically modern humans have co-existed with other primates for tens of millions of years."

The Out of Africa Hypothesis

In *Origin of Humankind*, Richard Leakey maintains that "the evolutionary activity giving rise to modern humans took place in the interval between half a million years ago and 34,000 years ago." He writes that "ripples of evolution" were going on in many different populations throughout the Old World during this period, culminating in a varying anatomy labeled "archaic *sapiens*." The concept of "ripples of evolution" might perhaps be better described as gene flow from a "revolutionary event." Let's see if Leakey's theories coincide with this suggestion.

Leakey describes the evolutionary model called the Multiregional hypothesis. This view posits that the origin of modern humans was a phenomenon which encompassed the entire Old World wherever populations of *Homo erectus* had become established. Leakey cautions against this approach, specifically in light of the emergence of new dating methods called electron spin resonance and thermoluminescence. Using these new methods, researchers have overturned the neat sequence of events which had earlier described the evolution of Neanderthal to modern *sapiens* in caves in Israel. These new methods suggested that the human fossils from Skhul and Qafzeh were actually older than most of the Neanderthal fossils by as much as 40,000 years, suggesting that Neanderthals were *not* ancestors of modern humans. How can this curious enigma be explained?

An alternative hypothesis asserts that modern humans arose in a single geographical location—most likely sub-Saharan Africa—and replaced the existing pre-modern populations after extensive and rather sudden migration into the Old World. Interestingly, this model has been variously called "Noah's Ark" and the "Garden of Eden" hypothesis, but most recently has been termed the "Out of Africa" hypothesis. Leakey points out that in this model "these populations would have shallow genetic roots, all having derived from the single, recently evolved population in Africa."

If it were true, the Out of Africa model would predict that the fossil record would not exhibit maintenance of regional continuity over time, but would show a displacement of regional characteristics with modern African characteristics. This is mainly true with regard to physical stature; indeed, the stocky, short-limbed Neanderthals seem to have been entirely displaced by the tall, slightly built, long-limbed people. Furthermore, if modern humans had emerged more or less simultaneously throughout the Old World, this would be evidenced by the fossil record. Clearly, it is not.

The oldest *modern sapiens* fossils—dated approximately 100,000 years before present—are limited to northern Africa, the Middle East, sub-Saharan Africa, and Israel. Leakey writes: "no fossils of modern humans of this age have been found anywhere else in the rest of Asia or Europe." Keep in mind Leakey's other statements that "no unequivocally identified *Homo erectus* fossils have been found in Europe" and that human fossils from Israel are older than most of the Neanderthal fossils by as much as 40,000 years.

The majority of population geneticists believe the Out of Africa hypothesis to be biologically plausible, and are skeptical toward the Multiregional model, which requires extensive gene flow across large populations, over a large geographical area, and over a very long period of time. These populations would have to be genetically linked while at the same time allowing for evolutionary change toward modern human characteristics over the entire, dispersed population. This is unrealistic given that such a scenario would, in all probability, produce more geographical variation than is actually seen.

In terms of tools and art objects, how are modern humans recognized in the archaeological record? The fossil record shows an increased complexity in simple stone tools about 1.4 million years ago; a gradual change from what is called the Oldowan to the Acheulian stone tool culture. Anthropologists use the Acheulian implements to identify the so-called *Homo erectus* populations. While this increase in complexity is notable, it changed very little over a very long period of time. Leakey writes that stasis, not innovation, characterized this era.

When change came, however, it was, in Leakey's words, "dazzling." The Upper Paleolithic Revolution, he writes, was "so dazzling that we should be aware that we might be *blind to the reality behind it*." Beginning about 35,000 years ago in Europe, fine tools were made from bone and antler, and tool kits comprised more than 100 different implements, including tools for engraving, sculpting, and making clothing. This inexplicable revolution exhibited life-like animal carvings, beads and pendants, and cave paintings as part of an innovative culture which appeared rather abruptly across the Old World. Leakey writes: "unlike previous eras, when stasis dominated, innovation is now the essence of culture, with change being measured in millennia rather than hundreds of millennia."

The Mitochondrial Eve Hypothesis

According to Sumerian records, the operation carried out by the Elohim on the female Primitive Workers resulted in Adam and Eve discovering their sexuality, or "knowing"—the biblical term for sexual procreation—and became "as one of us:" able to give birth or, perhaps more specifically, genetically compatible. The timing of the genetic manipulations are of particular interest. Sitchin asserts that the Sumerian record places the first genetic manipulation at about 300,000 years ago, and the second at about 250,000 years ago. Biologists now subscribe to the theory that there was an "Eve" in southeast Africa about 200,000 to 300,000 years ago. The *New York Times* reported in November of 1995 that scientists have concluded there was an "Adam" about 270,000 years ago, which, Sitchin suggested in an interview, is "exactly, give or take a day or two, the date I propose based on Sumerian writings."

The Mitochondrial Eve hypothesis is supported by research which traces the mitochondrial DNA inherited solely through the maternal line. The reason that the organelle mitochondria is traceable through the maternal line is that the only mitochondria which becomes part of the cells of a newly formed embryo are from the egg and not the sperm. Thus, scientists have traced the genetic ancestry of humans to a female who lived in Africa about 150,000 years ago, who was "part of a population of as many as 10,000 individuals." Leakey writes in *Origin of Humankind* that earlier dates have since been established; presumably dates closer to those given in the *New York Times* article mentioned above.

The Mitochondrial Eve hypothesis essentially supports the Out of Africa model, with one exception. The Africa model presumes that the expansion of modern African populations included some interbreeding with Old World pre-modern populations. The Mitochondrial Eve hypothesis, however, *does not allow for any genetic interbreeding with Old World populations*. It asserts that the modern African populations which spread over the Old World completely replaced existing populations, with interbreeding being almost a non-occurrence.

Analysis of present day human mitochondrial DNA reveals no evidence of interbreeding with other pre-modern populations. Of four thousand samples of mitochondrial DNA taken from people all over the world for this project, no "ancient" mitochondrial DNA has been found. All samples are remarkably similar and indicate the fairly recent and common origin of human beings. If genetic mixing had occurred between ancient populations, some people would have different mitochondrial DNA. Therefore, Leakey charges, the "modern newcomers" appear to have completely replaced the ancient populations between 150,000 to 50,000 years ago (give or take). The most disconcerting element of this theory seems to be the fact of the exclusion of genetic input from other populations which apparently co-existed. How did such a total replacement occur? Was it a violent genocide?

The disappearance of the Neanderthal population from the Earth is a mystery. In his book, *The Last Neanderthal*, Ian Tattersall suggests the violent demise of their dwindled population at the hands of other *sapiens* groups. Richard Leakey claims that there is no evidence for a violent replacement of species, and he allows for the occurrence of some interbreeding, even though it may not be reflected in the archeological record. Tattersall indeed allows for this conjecture as well. But strangely, there is an explanation which mainstream conjectures do not take into account. Did the Neanderthals die in a planet-wide catastrophic incident which the uniformitarian-based geological column utterly denies?

Who Are the Neanderthals?

The species *Homo neanderthalensis* was more than simply a variation of ourselves, writes Tattersall in *The Last Neanderthal.* On the contrary, the "big-brained Neanderthals," with larger brain cases (1500 ml.) than the current worldwide average (1400 ml.), were highly successful during a period of extremely tough climatic conditions in Europe and Western Asia for 150,000 years, and perhaps longer. Tattersall adds "as far as we can tell, that's a good deal longer than our own species has been around." It should be noted that Neanderthals are not necessarily believed to be a different species, but a variation of *sapiens*.

It is generally agreed that different varieties of humans were widespread in the Old World during the time of the dispersal of the Neanderthals over Europe. Classic Neanderthal-type fossils are a more or less local phenomenon spread sparsely over Europe, with a higher percentage of finds in western France. To date, however, there is no biological evidence that the Classic Neanderthal type ever occupied Africa, Arabia or Asia. However, according to *Hidden History*, Louis B. Leakey once suggested Neanderthals and other variants were the result

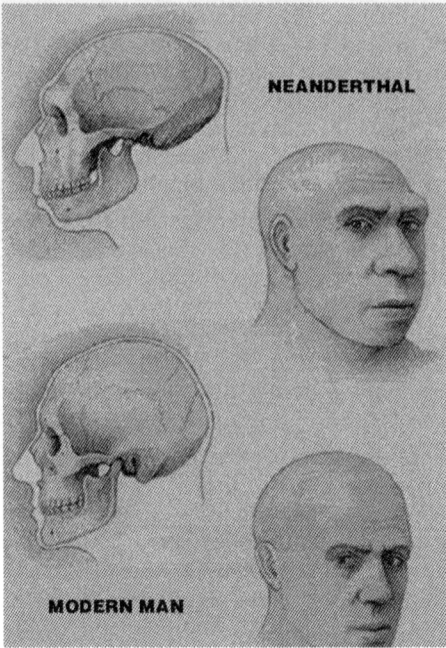

NEANDERTHAL

MODERN MAN

Neanderthals were much more modern than previously thought.

of crossbreeding between *sapiens* and *erectus*. He indicated no problem with the ability of these two forms to interbreed, which suggests he considered them to be the same species.

The theory is widely peddled that the Western European Neanderthal population "had a range of vocalization too limited for the development of proper language." Some believe *Homo neanderthalensis* to be a separate species from *Homo sapiens*, while others believe it is a subgroup which could interbreed with modern humans and, it is suspected, "would have done so if both were present in the same place at the same time." However, the *Physical Anthropology* text book presents the erroneous statement that Neanderthal populations of Europe probably did not have a modern vocal tract, even though the brain case was large and modern in respect to parts important for speech. This information has since been contradicted.

According to Jack Cuozzo, in *Buried Alive*, a Neanderthal skeleton found on Mt. Carmel in 1983 was discovered alongside modern stone tools. The Kebara II fossil had the only hyoid bone ever found in a Neanderthal. The hyoid bone, a floating bone in the neck, is essential for speech. The Kebara II Neanderthal proved that Neanderthals did indeed speak. It is probable that the Neanderthals are much more modern than the current theory allows.

The Sapiens Sapiens Signal

Although a "dramatic event" is recorded in Western Europe about 35,000 years ago, Leakey and others believe that this does not necessarily indicate that the "final emergence" of modern humans occurred in Western Europe, as earlier believed. Paleontologists now generally recognize Western Europe as "something of a backwater" on the foreground of an advance that swept from *East to West* across Europe. In this view, the Neanderthal populations "disappeared and were replaced by modern humans." Importantly, Leakey believes the Upper Paleolithic Revolution in Europe was a "demographic signal and not an evolutionary signal."

From where did this demographic signal emanate? Leakey believes, on the basis of fossil evidence, "Africa, in all probability—or perhaps the Middle East." Leakey qualifies that, despite the scarcity of technological evidence, since a narrow blade technology does appear in Africa about 100,000 years ago, we have to go with the African origin of modern humans. Supported by an abundance of fossil evidence in the Middle East region along with new dating methods, it is also probable that Neanderthals and modern humans essentially co-existed in the Middle East for as along as 60,000 years. Some Neanderthals of Europe have been dated at 100,000 years, making them contemporaries of modern humans in the Middle East.

The Extraterrestrial Hypothesis can provide an explanation for "anomalous" data exhibited by the fossil record. According to the Out of Africa model, mankind seems to have emerged from both Africa and the Middle East, the cultural nests of the Anunnaki during their incumbency on Earth. The Out of Africa model can be seen in the context of the mining operations which began there approximately 300,000 years ago, culminating in the cloning of a human work force. This is the "*sapiens sapiens* signal," the "dazzling" cultural achievements of a certain core group of humans which were moved from Africa to Western Europe, virtually overnight.

According to Richard Leakey, the known *sapiens sapiens* signal in Western Europe is far richer than in Africa. For every archeological site of this era in Africa, there are about two hundred such sites in Western Europe. Leakey maintains that this disparity reflects "a difference in the intensity of scientific exploration in the two continents, not on the reality of human prehistory." The Extraterrestrial Hypothesis suggests that this disparity *may actually* reflect the "reality of human prehistory." We must follow the *sapiens sapiens* signal.

The Sumerians have written that, after a time, some of the hybrid miners remained working the mines in South Africa and others were transferred back to Mesopotamia, the location of the biblical Eden. It is here that the

The Sapiens Sapiens Signal. Cave art from Lascaux, France depicting horses and bison.

Anunnaki had first landed and began their irrigation and building projects, and where they subsequently set up cultural centers for the propagation of human knowledge. Thus, although Africa is the "birthplace of humankind," the Sumer region is the "cradle of civilization," which began with transplants from the core Africa group.

The story is relayed in the Sumerian texts. According to Sitchin in *Wars of Gods and Men*, as the population of the Primitive Worker increased, gold production in Africa increased as well. This in turn caused stress on the Eastern front, in the refining and smelting centers of Mesopotamia. In addition, the waters in this region were constantly overflowing, and dikes and canals were constantly in need of being dug. The Anunnaki workers in Mesopotamia soon began asking for their own Primitive Workers. They made a request for "the Black-headed people to give the pickax to hold." A text called *The Myth of the Pickax* chronicles the denial of transfer of some of the Black-headed people to the Mesopotamian region, and the subsequent heavy artillery attack on the walls of the central compound in Africa where they were being protected. As Enlil's weapons broke through the fortification, the Primitive Workers spilled out of their compound. It is written that Enlil "eyed the Black-headed Ones in fascination."

Ancient drawings thereafter depict the Primitive Worker toiling for the gods in Mesopotamia, building their houses, digging canals, growing food, rowing in boats, and waiting on the gods. During this "animal like stage in human development," archaic *Homo sapiens* performed their work as "naked as the animals of the field." It is written that, when mankind was first created, they "ate plants with their mouth like sheep, drank water from the ditch." Were these people our "Neanderthals"?

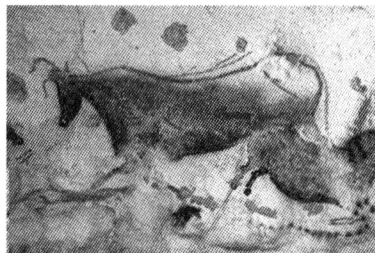

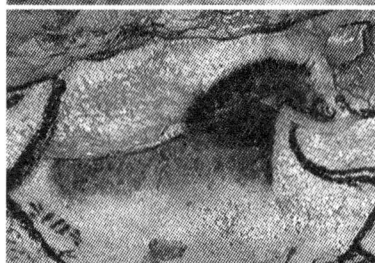

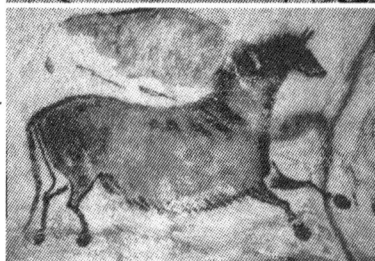

The various layers of cultural indoctrinations taking place in or near cultural centers over a long span of time, which are part and parcel of the Extraterrestrial Hypothesis, would explain the apparent cultural disparities noted over geographical areas once some of these people left cultural centers in the Middle East. The Extraterrestrial Hypothesis would explain the apparent co-existence of primitive stone tool sites in some areas of the globe and high cultures in other areas of the globe between 10,000-30,000 years B.P. What else but the existence of an "anomalous" cultural factor, introduced from the outside, could adequately explain this incredible discrepancy in cultural "evolution"? *OMNI Magazine* published a story in January, 1995 about Samsat, a city with a population of 50,000 along the Euphrates River Valley during the Roman Empire. The history of Samsat dates as far back as the *Neolithic Period*. Tablets found in the area written in cuneiform, a system of writing developed by the Sumerians, is proof of the spread of the Sumerian culture into Turkey.

Four such sites which record the history of the Sumerians in the area "may contain artifacts that overturn conventional notions of how and where civilization began." Another city called Kazane Hoyuk, had a population "unheard of *before* the development of agriculture and civilization." This prehistoric city was one of the "independent seeds of civilization" which "nurtured cultural advances at about the same point in history." Thus, while some humans on the globe were working with basic stone tools, others were in the midst of a "dazzling" cultural revolution. In addition, strange similarities have been noted between distant cultures which are not known to have been in contact. These anomalies remain unexplained in archaeology, and are explainable only by a "revolutionary event." Let's take the Mayan long count system as an example.

The Maya Long Count and Space-Time/Movement

Ancient cultures were inexhaustibly concerned with counting cycles of time regarding the planetary bodies of the solar system. As Charles William Johnson has noted ("Earth/matriX: Science in Ancient Artwork," at www.earthmatrix.com), by counting time, one necessarily is counting space as well, for reality exists as space-time, and space-time is movement.

Johnson writes that the numbers of the "Maya long count" are notoriously long and may be related to universal numbers pertaining to such things as the velocity of the Earth, the Sun, and the speed of light. He notes that the reckoning of time in the Maya long count system seems to reflect cycles of time that are "irrelevant to an individual's lifetime on Earth." The meaning of these long numbers do not appear to have meaning from an Earthling point of view. It has been suggested that the Maya long count was a reckoning of relationships of the planetary bodies within the solar system, with particular interest in the relationships of Venus and Earth, the Sun and the moon, and the opposition of the Earth and Mars.

Johnson's mathematical analyses suggest that the ancient Mesoamerican systems and the ancient kemi or Egyptian system together represent a single system. He posits that all of the day-counts "emanated from a single mathematical model of theoretical conception, based on astronomical observation of the cosmos." These reckonings have been observed by modern scholars as "errors in observation of the ancient astronomers, which were later corrected." However, Johnson disagrees. He writes:

> From the extreme coincidence of numbers it is possible to consider the fact that the Maya long count numbers/fractals represent a system which may have taken into consideration the spatial and relational aspects of time and its measurement. The reckoning system of ancient Mesoamerica would appear to represent a single system of reasoning, supporting the idea that it appears to have been developed at a single stroke.

At the same time, another enigma presents itself. Johnson maintains that this enormously complex number system appears to have "employed methods of analysis (duplation/mediatio) which were worked out over thousands of years of human endeavour given its apparent simplicity for managing complex numbers." Thus, we have here a system which appears to have been devised "at a single stroke," or all at once, but which is so complex as to have taken millennia for human beings to develop. In addition, according to analyses of the numbers in relation to the concept of time-cycles, Johnson maintains, "it would appear that the purpose of the system was so abstract as to defy any significant meaning." This is because the abstract relations between planetary bodies were being computed from the perspective of space (bodies) and movement (distance traveled). These ancient astronomers, Johnson suggests, were measuring the exact distance traveled by the planetary bodies throughout the Universe, or at the least, within this galaxy.

But how did the ancient astronomers know that planetary bodies travel in courses which can be designated in numbers? In considering whether this knowledge was imparted by a visiting culture, Johnson wonders why they wouldn't have been shown "a more streamlined manner" of annotating the computations, rather than the laborious dot-dash system. He notes that the manner in which the numbers are situated and the data is generated seems "far too elaborate and even clumsy."

Nonetheless, Johnson suggests that there appears to be a "conspiracy of numbers" between the calendrical system and space-time events. Significant numbers of the Mesoamerican reckoning system reflect the time cycles of astronomical events as well as the distances traveled by those events. He writes: "The numerical progression chosen by the Maya for their long count system (360c) just happens to reflect a similar progression found in the space-time events of the solar system and the Universe, not only with respect to space-time as such, but with regard to movement (velocities)."

He suggests, to maintain that the calendrical computations were made in error, then, is "to consider the entire system to be an elaborate structure of errors." On the contrary, he writes, the coincidences appear to be due to design. He writes: "The fact that particular numbers were assigned distinct meanings would not appear to have been due to error in computation, nor to the simple logistics of numbers by themselves, but rather that the numbers are reflecting the measurement of space-time/movement as it exists, as we also know it."

Certain known equations of the ancient reckoning system appear to reflect at least a sectional view of " space-time coordinates." The ancient reckoning system of ancient Mesoamerica and ancient Egypt are far too elaborate and precise to believe that they are elaborate systems of error. He suggests, given that "space-time appears to exist in progressions of relations, fractal expressions, it might not be too difficult to think that the numerical progressions alone might suggest certain possible answers to comprehending the next level of space-time existence."

Johnson also notes a correspondence between the languages of ancient Mesoamerica and ancient Egypt. Such a correspondence has been denied, since it is argued that these cultures never met. He writes, there is no historical data to link the peoples of ancient Mesoamerica and the peoples of ancient Egypt. Yet, there are obvious linguistic similarities between *nahuatl* and the ancient Egyptian language. He writes:

> Linguistic correspondence between *nahuatl* and ancient Egyptian appears to represent a smoking gun; that is, a trace of evidence that these two peoples did enjoy some kind of contact between themselves ages ago. The fact that we have no real evidence of said contact, or that we have been unable to find any

such evidence, should not serve as the basis for denying the possibility of that contact. To attribute all of these similarities in sound, symbol and meaning to mere happenstance seems to be a very unscientific way of resolving an annoying issue.

In addition, Johnson has studied in detail the ancient artwork known as "Pakal: The Maya Astronaut." In his work, displayed on the cover of this book, he has duplicated this ancient artwork from various perspectives, and analyzed the various parts of the vehicle in which the "ancient astronaut" is sitting. By rotating certain sections of the vehicle according to the direction implied in the artwork itself, he maintains, the vehicle takes on "a logic of its own." It would appear, he suggests, that the artist had conceptualized some type of spacecraft. The images that result are highly suggestive of "lift-off" and space travel.

As far as we know, the Maya did not have the technology for space travel, Johnson writes. If they did, one would think that at least a piece of such a vehicle would have survived somewhere. Therefore, the Pakal stone sculpture may represent a visit by space travelers. He suggests that homage may have been paid by rendering this depiction in stone. The manner in which Pakal is sculpted is characteristic of "a profound and exact scientific knowledge, resting on astronomy, mathematics, geometry, mineralogy, all of which was expressed in their artwork."

The precision reflected in this artwork, Johnson writes, reveals the nature of the knowledge that they were attempting to communicate. It was sculpted in stone for all to see, possibly for generations to come. In fact, he writes, the knowledge was registered in such a manner as to permit learning it and knowing its intricacies. Scientific knowledge was encoded into the sculptures and the architecture in such a manner, and with such precision, that anyone who studies it will discover its logic. In this sense, Pakal the Ancient Astronaut can be seen as a "time capsule."

Does Pakal the Ancient Astronaut represent an artifact of proof that the ancient peoples may have been visited by spacefaring ETI? Johnson sees such visits, if they did occur, as simply "encounters among different peoples who populate this Universe together; each learning from the other." He writes:

It should not bother us that other beings have existed or exist in this averse immense Universe. Nor is it difficult to see why we might think we are the only children of God after having lived here for so long on this seemingly isolated planet in this apparently isolated solar system. Nor should it bother us that possibly we were once or twice visited by others from this vast Universe, if they came in peace and the ancestors of the Earth possibly saw fit to honor them in their artwork.

Thus, the Extraterrestrial Hypothesis can explain the most puzzling revolutionary event: how an animal became intelligent in so short a time; an animal whose brain seems quite capable of *accepting* culture (being taught), but not of *devising* it. It is clear that some of the anomalies strewn in the path in the search for human origins are consistent with the extraterrestrial genetic engineering scenario. First and foremost, however, those opposed to the idea of space travelers need to put their preconceived ideas aside in order to look into the possibility. As previously discussed, have we put the cart before the horse with respect to the assertion that space travelers cannot exist because *we* exist?

The Extraterrestrial Hypothesis might also explain a phenomenon anthropologists have chosen to call "simultaneous novelty" to describe the curious similarities between cultural achievements of highly civilized states found in distant areas of the globe. The Sumerians report that a group of humans departed with one of the Anunnaki leaders to South America where they set up a cultural center. Incredibly, some South American stories describe their arrival to the New World in underwater ships, as well as metallic "birds." In addition, the peopling of Australia circa 60,000 years ago necessitated at least a 60-mile open ocean journey. How did they traverse the globe so long ago?

The Sumerians were consummate shipbuilders. They even gave mankind explicit instructions on the building of a totally waterproof ark in order to save himself from a coming Earth disaster. Which brings us to the explanation for one of the most puzzling problems of the fossil record: how the Neanderthal populations disappeared from the Earth and were replaced by anatomically modern humans.

Universal Deluge Stories

Just before the deluge, Sitchin reports in *Wars of Gods and Men*, a new Ice Age began, resulting in much suffering, starvation and even cannibalism. Enlil, the leader in charge of the Earth colony, wasn't happy with the situation. After imparting the Knowing to the hybrids, mankind's days of innocence were gone. They now needed to wear garments for they knew shame and sexual lust. It is suggested by drawings of Enki in tethers that he was arrested after this unauthorized deed, and Enlil ordered the expulsion of The Adam from the E.DIN. No longer confined to Anunnaki settlements, mankind was free to make the Earth his home.

Even the *Book of Genesis* records the sexual relationships which began to occur between "the sons of the gods and the daughters of men" once the gods saw that "they were compatible." Disturbed by the "growing togetherness" of the young Anunnaki and the descendents of the Primitive Worker, of mankind's increasing obsession with sex, and as well by the suffering caused by the food shortages of a new Ice Age, Enlil decided to wipe mankind off the face of the Earth. As it turned out, he didn't have to do much besides refuse to lend a hand to the victims of a natural catastrophe.

The story of a universal flood is told by all ancient civilizations, in all oral traditions, in all written systems, on all continents, and on all islands, all over the world, "even in places never visited by missionaries," wrote Velikovsky in his unpublished manuscript entitled *In the Beginning*. (www.velikovsky.collision.org). He noted that rabbinical sources refer to the building of many arks before the deluge. Velikovsky suggested that there were perhaps many Noahs, but only one of the arks actually survived. He wrote, however, that ancient stories say that other people survived the flood waters, in caves high in the mountains, in far separated regions of the Earth. Like the former catastrophe of the Fall of Man, the deluge changed the nature of herb, animal and man. The continents changed their places, and even the sky was not the same.

The astrological connection between Saturn and water catastrophes has a very ancient origin, wrote Velikovsky. Many ancient astrologers, including Ptolemy, attributed floods to the planet Saturn. The Hindus assigned the deluge to the end of the Satya yuga and the reign of Satyavrata, or Saturn. The Saturnian age was the age of the flood, and Brahma (the planet Saturn) is said to have warned Manu of the deluge. After the waters covered the Earth, Brahma is said to have "floated over the expanse of the ocean." An ancient woodcut portrays Brahma seated "on a rayed disk," which is hovering over the waters of the deluge. The woodcut contains the words: "then the lord floated over the vast ocean, void of the sun and Moon."

In his publication, *Kronos*, Velikovsky noted that the identification of Brahma with Saturn is evidenced by the fact that the god is assigned a celestial sphere. A celestial sphere could be interpreted as an "orbit." Velikovsky noted that legends tell that the "high-souled Brahma is seated in the highest abode," and that the "highest celestial sphere is that of Brahma." It is written in the

Vishnu Purana that Brahmaloka, the heaven of Brahma, is the "seventh and highest heaven."

Velikovsky ascribed the universal deluge to a catastrophe involving the planet Saturn, which, he believed, exploded like a supernova. The date of the event, he claimed, was possibly about ten thousand years ago. He believed the solar system, and perhaps points beyond, were illuminated by this exploding star, "and in a matter of a week the Earth was enveloped in waters of Saturnian origin."

Sitchin relays the account from the Sumerian point of view. As he relays it, the scientific monitoring station situated in Africa began to report that the Antarctic ice cap was becoming unstable, since the dense ice cap was resting upon a layer of slush. In addition, just as this situation was duly noted, the home planet Nibiru was about to make its 3,600 year appearance in the Earth skies, and its gravitational pull could likely cause the precarious ice cap to come crashing into the sea. The Igigi stationed on the orbiting platform above Earth confirmed the impending catastrophe and an Assembly of Gods convened. Enlil made all of them swear to keep the coming deluge a secret from mankind, since Enlil decided he wanted the humans gone from the Earth. However, Enki disobeyed and gave very specific instructions to Ziusudra (the Biblical Noah) to build a completely submersible vessel.

As Sitchin reports in *Wars of Gods and Men*, Sumerian texts state that, from their shuttlecraft above the Earth, the Anunnaki watched the waters of the deluge destroy everything they had built. Giant tidal waves pounded the Earth mercilessly until "all that had been created, turned back to clay." As the waters subsided, the Anunnaki landed on the first available area of dry land—Mount Ararat. Noah and his entourage arrived shortly thereafter. Enlil was outraged, but he relented after being convinced that they needed the Earthlings to help them start over again.

If the deluge story is true, we would find the emergence of the earliest civilizations in the highlands, the places where global waters would first recede. This is exactly what the archaeological record shows. Archaeologists have always been curious about the peculiar origins of husbandry and agriculture in the mountainous regions of Mount Ararat, a strange place for agricultural beginnings. Even the *Table of Nations* in *Genesis 10* seems to suggest that higher areas were populated first. Sitchin illustrates that the *Table of Nations* provides an accurate archaeological record of the first settlements of the Earth after the deluge. The first seven nations to be established by the lineage of Noah were situated in the highlands of Asia Minor, the Black Sea and the Caspian Sea. The lower coastal areas and islands were settled by much later descendants, since these areas were not habitable until much later. In addition, the descendants of Ham ("The Dark Hued One") settled the African nations, beginning with the mountainous regions. Descendants of Shem settled the highlands of the Persian Gulf where the spaceport had once been and where a new mission control center was built.

Neanderthals and the Great Flood

In their book, *When the Earth Nearly Died*, D.S. Allen and J.B. Delair provide a detailed overview of certain peculiarly situated remains which have been found in caves and rock fissures all over the world. In what they call "subterranean charnel houses" are found the remains of "normally incompatible kinds of animals lying in unnaturally close juxtaposition." Often jumbled up with large stones or sizable boulders, and enveloped in mud, Earth, or iron-ore deposits, the broken remains of these animals have been "deposited in great confusion," and obviously under extremely violent conditions. For instance, a near complete rhinoceros was discovered in Derbyshire, England, lying in a "vein of lead." The authors suggest these veins of lead may indicate an ancient cosmic collision with an asteroid or a "heavy" celestial body.

Also discovered lying in peculiar disarray in caves in France were the remains of lion, bear, wolf, fox, giant bison or ox, horse, rhinoceros and other animals. Viewed collectively, the assemblage of animals found all over the world exhibit "an extraordinary range of habitats and climatic conditions, and resemble the mixed faunal assemblages noted all over Europe." In these caves and rock fissures are found the jumbled masses of "jungle" animals, "tundra" animals, "prairie" animals, "desert" animals, "marine" animals, and people. A cave in Monaco includes the bones of lions, rhinoceros, hyenas, monkeys, elephants, and whales. By what strange Earthly perturbation could the remains of large marine creatures end up in a cave in France?

The distribution of this phenomenon on nearly all continents indicates that its cause cannot have resulted from the slow movement of ice postulated by uniformitarian glacial theory. Just as unlikely is the commonly touted "quicksand" explanation of the quick burial of almost complete Mastodon carcasses on the North American continent.

In short, the fossil record illustrates that in fairly recent times these worldwide biological decimations were catastrophically sudden. Many organisms "perished literally where they stood," and some were buried in erect or walking positions. According to *When the Earth Nearly Died,* there is no other scheme which comprehensively accounts for the anomalous distribution of typically "Miocene" and "Pliocene" species alongside the supposedly younger "Pleistocene" species, and the geographically discontinuous distribution of animal remains. These worldwide discoveries of climatically incompatible species, modern and ancient forms, or forms which no longer flourish at the latitudes where their remains now occur, "loudly proclaim a sudden and violent extinction."

The most enigmatic disclosure is the following: several caves near Santa Lucia contain human bones intermixed with those of cave jaguar, horses, llamas, and other animals. One cave contained the bones of over fifty humans of all ages, buried in hard clay and mixed with the bones of an assortment of late Pleistocene animals. The skulls of the ancient humans were of the old European Neanderthal type. It has otherwise been argued that the Neanderthals died off between 25,000-30,000 years ago, and that their displacement by more modern humans may have been violent. Although it is likely that their population had dwindled significantly, this evidence clearly suggests that the Neanderthals did indeed survive alongside modern humans until approximately 11,500 years ago, or 9,500 B.C., and that they indeed died violently, although not at the hands of modern humans.

Thus, the reason for the disappearance of the Neanderthals can be explained by the Sumerian story of the great flood, told by hundreds of cultures worldwide, from which only modern humans were saved by specific instructions imparted to one family to build a totally waterproof, submersible ship. All other populations would presumably have died out in this catastrophic event, except for the DNA of the animals which were deliberately put on board the ship by an Anunnaki leader "god." (See Sitchin's *Genesis Revisited*)

The biblical ark of Noah is identical in concept with the Greek ark of Deukalion. The ark of Noah is also the same as the ark of Ziusudra or the ark of Utnapishtim in the Middle East. The story of the flood and the ark, which contained the "archetypes" of all living creatures in pairs [the word *arche* in Greek is related to *ark*], goes back to at least Sumerian times. Robert Temple writes in

The Sirius Mystery that the ark stories of the Greeks and of the Hebrews are "extremely late forms of an exceedingly ancient story, which existed thousands of years before there were such things as Greeks or Hebrews in existence." His research concludes that both sources are describing "magical ships in which sit 'those who come out of the womb,' in the sense that they repopulate the world after the deluge." Specifically, the reference to those who come out of the womb, Temple writes, "seems to refer to the children of the Earth goddess springing from the womb of the Earth." In Sumerian, the word *ki* means *Earth*, but also refers to Nintu, the goddess who gives birth. (see also Gordon, *Common Background of Greek and Hebrew Civilizations*)

Are Neanderthals Post-Flood People?

Returning to our discussion of the enigma of the Neanderthals, indeed, some humans with "Neanderthal" features have been noted to be contemporaneous with humans exhibiting more "modern-looking" features. But it is important to stress that these differences could be due to sex, age at death, and general differences in genetic traits. In addition, given the earlier information regarding the subjective application of dating methods, we should always keep in mind that these dates could be way off.

A new theory of Neanderthal lineage has been presented by Jack Cuozzo, in his book *Buried Alive: The Startling Truth About Neanderthal Man*. Cuozzo, an orthodontist, makes several convincing arguments, as well as a few not so convincing assumptions based on Biblical parallels. I should note that the reason I purchased Cuozzo's book was because the title suggested I would find more information on the type of quick and violent Neanderthal cave burials I had read about elsewhere. But, Cuozzo makes the strong assertion that the Neanderthals he has been allowed to study "all appear to have been post-flood people, mostly buried by relatives and friends." He argues: "The Neanderthals are post-flood people, not square one people." Therefore, I am perplexed at the title of this book, but nonetheless find the theories interesting.

The hostile climate of post-flood Europe, with its frigid climate, rugged terrain, and proximity to glacial ice sheets, stressed mankind's "toolness" to its fullest. However, we are not to think of Neanderthals as "troglodytes," or dumb cave dwellers. Instead, Cuozzo urges us to realize that these are intelligent people who chose a nomadic existence instead of staying on in the Middle East. For some reason, he writes, they continued to migrate away from the mountains of present-day Iraq. Caves provided shelter from cold, harsh climatic conditions.

One of the major points Cuozzo makes is that Neanderthal prehistory is "made to look like 'square one' by all the museums, or perhaps square two or three if you take into account *Homo erectus* and the Southern apes." But, Cuozzo insists, the Neanderthals were not square one, but were survivors of the Great Flood. He writes, "the Earth that their fathers had known was gone and newly created post-flood conditions prevailed. Because of this fact, uniform thinking is both inadequate and dead in this time frame of history."

This is an important, two-part statement, from which we can deduce: (1) the flood changed everything in the stratigraphic layers, since it is a major upheaval of the earth, and (2) the conditions following the flood forced advanced, intelligent people to move into survival mode. This means that the confusion in toolmaking styles lies in the fact that, at this point in time, everything away from the Middle Eastern centers was reverted to earlier toolmaking styles out of necessity. The term "Neanderthal" shouldn't bring up the image of a "troglodyte," but the image of an advanced human group caught in a very difficult survival situation. In this sense, they are not "square one."

It is time we realize that there are, indeed, many other time frames in our prehistory where uniform thinking is inadequate. It is impossible to subtract catastrophic theory from the seemingly "incremental" achievements of our ancestors.

In addition, there may not have existed any large groups of Neanderthals at all. Cuozzo points out that, as a group, there have been more Neanderthals discovered than any other group, yet "the total number of fairly complete Neanderthal skeletons is only about a dozen." As Cuozzo writes, it's easy to portray the myth of "groups" of individuals when paleontologists count like they do. Anthropologists portray this myth in two ways: by exaggerating the number of individuals actually discovered at sites, and by calling their toolmaking ability an "industry." An "individual" can be a piece of jaw bone, a fragment of skull or bone, a single tooth, or a jaw section with a few teeth. Each is usually considered as deriving from a different individual. For instance, in one Neanderthal site in Monsempron, France, Cuozzo illustrates that the evidence for what has been described as seven individuals can be read as two or three individuals at most. Another site which Cuozzo describes, in LaQuina, France, exhibits what could be the remains of six or seven individuals, rather than the 21-22 listed in the record. These types of exaggerations are applied across the board for all human fossil finds.

Most people are unaware that these fossils are locked away in museums, away from the professional scrutiny of those who would potentially correct the textbooks, and especially if those people happen to be creationists. As Cuozzo explains:

> No person with a creationist worldview, to the best of my knowledge, since Darwin's people took over science, has ever penetrated behind the evolutionist's lines to study their fossils with x-ray equipment. The fact that some paleoanthropologists consider them to be their fossils and not science's fossils, or open to experienced workers in the field, has been well established.

Cuozzo writes that people with Neanderthal-like features have been found in areas as diverse as China, Iraq, Belgium, Germany, France, the Isle of Gibraltar, and even South Africa. The common dates attributed to the Classic Neanderthal type in Europe is 130,000 to 35,000 years B.P. During this span of time, which is equal to approximately 10,000 generations, Cuozzo argues, we should find more than a dozen partially complete skeletons. The picture we are provided by museum reality engineers is an elaborate mythology. In this regard, Cuozzo writes that paleontologists are able to make up almost any story they want about the "temporal relationship of the layers of the Earth and debris found scattered around the world." Archeologist Wymer has admitted that "there are no rigid rules and certainly a subjective element is introduced, for the archeologist must take each case on its own merits." (Cuozzo, 108)

What does one do when finding something too archaic in a shallow stratigraphic layer? According to Cuozzo in *Buried Alive*, just say it came from a lower level. For instance, he writes, a Japanese team digging in a cave near the Sea of Galilee discovered a Neanderthal skeleton in a layer associated with Upper Paleolithic artifacts as well as pottery. The date range in this stratigraphic layer was 27,000-28,000 to 5,710 years B.P. With the latter date being considered "too young" to fit the pre-ordained paradigm for Neanderthal finds, the team blamed carbon contamination from an upper level. The date they eventually settled on was 40,000-50,000 years B.P.

Face: The Facts

Cuozzo, an orthodontist, has deduced from his studies of Neanderthal head, face, jaws and teeth that Neanderthal people were very long-lived humans. He also points out that modern children are aging much faster than children of just a few hundred years ago. In terms of a comparison of modern children with ancient children, Cuozzo believes that the eruption of the first molar at age six would have been impossible in these children. To put this in perspective, Cuozzo surmises that the average age of sexual maturity in biblical times was about thirty years of age.

Based upon the fact that Neanderthals aged much more slowly and had a longer natural life span, Cuozzo suggests that Neanderthal fossils are largely misinterpreted. In addition, he points to age changes in the head and face, indicating that stronger features and a larger head and face might indicate that the person was older at burial. Studies have shown that the human head continues to grow in length and width well into adulthood and middle-age, and that the proportions change significantly. It has also been noted that there is a "sporadic thickening" of the skull in later life. It is a myth, Cuozzo asserts, that aging is synonymous with bone loss. Could the Neanderthal enigma be explained as simply "elderly" people, that is, *much* older people?

Average life expectancy has doubled from 40 to close to 80 years since the early 1800s. This is mainly because more children survive their early years of life than they did historically. However, the age at which a person becomes old in terms of natural infirmity, Cuozzo maintains, is exactly what it was in biblical times. This does not apply to the early peoples of Genesis, but to the majority of biblical peoples from Moses through the time of Christ. The early Genesis people are a different matter entirely. The Bible tells us that Adam lived to be 930, and Methuselah lived to be 969. Jared, the sixth generation from Adam, lived to be 962. Noah lived to be 950. The early patriarchs, Enosh, Seth, Kenan, Mahalel, and Lamech all lived life spans ranging between 800 and 900 years. But, we must ask, when did the clock start winding down?

Cuozzo's book addresses the real quandary presented by the fossil record: we find fossils with aged Neanderthal-like features at higher levels and later dates, while finding fossils with more modern features earlier in time. Along with this fact, we see the displacement of the stocky, short-limbed Neanderthals by taller, slightly built, long-limbed people. This peculiar set of facts has puzzled geologists for a long time, since it looks like Neanderthals were a "dead end," and it suggests they were not our ancestors.

Cuozzo maintains the Neanderthals were more vital humans and they lived much longer than the other human group with whom they co-existed; perhaps hundreds of years longer. He suggests that the Neanderthal people were the original people of Noah's lineage, vital people who lived much longer life spans than those who followed them in the next generations, due to rampant radiation on the Earth. The Neanderthal skeletons, Cuozzo suggests, are of the lineage of Shem, son of Noah, the men of Genesis 11. If we consider this hypothesis, the older men of Genesis 11, who lived to be about 400 years of age, would have been buried at higher levels, since they, ostensibly, outlived many of their offspring, including their great-great grandchildren. I fail to see how there could be such a dramatic change in the features of the immediately succeeding generations, *unless we presume two things*:

The aging effect in terms of head and face size is ongoing. In other words, does the head and face just keep right on growing into the third and fourth century? It seems like a stretch to consider that the Neanderthals look so different because they are elder patriarchal and matriarchal figures, but perhaps this is so. Perhaps their elaborate ritual burials should be seen in this light.

There was gene flow coming into the subsequent generations from another group, which would make their features more "modern" in comparison to the elders. If so, where did this group come from? Are we again bumping into the Biblical and Sumerian assertion that the "gods" procreated with the daughters of men? If we are talking about post-deluvial times, there are no other groups possible.

The picture that Cuozzo essentially paints is that modern man is not evolving, but *devolving*. While we are constantly told that the modern life span has markedly increased over the past few hundred years due to medical and scientific advances, we should not assume that the general trend over the long term has been an increase in the natural life span since ancient H*omo sapiens* or Neanderthals.

Cuozzo demonstrates that the reason human life is devolving is due to high radiation levels, which were particularly high just after the Great Flood. He indicates that this may be the reason for the great discrepancy in ages between the men of early Genesis, and those of Genesis 11; a full half-life reduction. Cuozzo suggests that the caves after the Great Flood would have had a high radon-222 count, as well as an accumulation of radium A (polonium-218), radium B (lead-214), radium C (bismuth-214), and radium D (lead-210). In the poor ventilation of the caves, radon-222 gas would coat the walls of the cave with lead-210 at high levels. Noah did not live in the caves, he lived in a tent, since he had been forewarned. However, he could not have entirely escaped the radiation which "spewed forth from the extensive volcanism and aerosols in the post-flood atmosphere." Noah's grapes would have been tainted with radium-226 and lead-210, and his tent had to be sitting on contaminated ground. (Cuozzo, 256)

Cuozzo points to a 1984 U.K. study which showed that beverages, cereals, vegetables, bread and sugars still contain radium-226 as well as lead-210. Yet, he notes, in a study which compared lead-210 from the walls of Mystery Cave in Minnesota to lead-210 on the basement walls of a house in Minneapolis, it was found that the lead-210 alpha activity in the cave was 100 times greater. The observed lowering of radiation levels and the increase in cancer, he posits, can be explained as a decline in immune response and protective genetic mechanisms, as well as an increase in other types of modern, man-made carcinogens. It is thought that the frequency of cancer was much lower in antiquity, and it is not known when the present high incidence of cancer began. Cuozzo suggests that ancient people may have had "better DNA repair mechanisms designed to detect replication or transcription errors." Background radiation now is very minor, except in caves, in radon-filled homes, and places like Chernobyl. Cuozzo states, "amounts of natural radiation figures seem to be decreasing as the Earth wears down."

Cuozzo has discovered in researching Neanderthal fossils that Neanderthal children exhibit a much slower growth pattern than do modern children. Cuozzo refers to this general physical trend as "immaturity." In comparison, moderns have lost this immature developmental tendency. A 1994 British study on the treatment of children with brain tumors and leukemia showed that high doses of cranial irradiation caused a more rapid onset of puberty. Early maturation is related to the overall process of loss of longevity, and the degeneration of mankind. The current "speeded up" trend in modern humans is the result of the degeneration of genetic material which, Cuozzo asserts, is due to the Biblical Fall of Man.

Cuozzo asserts, "after the Fall, all the harmful isotopes, rays, and gases must have been switched on when God said, 'cursed is the ground because of you.'" After all of the scientific work he does in this awesome book, the best Cuozzo can do is propose that God simply "switched" the clocks of radioactive isotopes and cursed the ground which his human creations walked upon, poisoning everything they needed to survive. According to Sumerian accounts, this is sort of what happened, but it wasn't simply a supernatural occurrence. As discussed in the previous chapter, the radioactivity on the Earth may *also* have to do with a purposeful nuclear genocide which occurred many generations *after* the Flood.

The Partnership of Man and Gods

Sumerian records indicate that after a time, and more so after the great deluge, the Anunnaki began to treat mankind as partners on Earth. During this partnership, mankind truly began to learn the tools of many useful trades and, as Leakey describes it, the *sapiens sapiens* signal is nothing short of dazzling. But remember, this cultural extravaganza was imparted in waves of indoctrinations which are recorded in the cultural record as "ripples of evolution." In reality, cultural change does not follow the straight lines and little boxes that some people like to draw.

In light of all of this newfound information, we might surmise that the Neanderthal populations of Europe may have been more archaic specimens of *Homo sapiens* which left the settlements of Mesopotamia before the "astronaut" sons began to sleep with the daughters of men in or near the urban settlements. Also, if we consider Cuozzo's work, we might speculate that the Neanderthals may have been the Biblical patriarchs and their family lineages, who followed commandments regarding sexual mores, thereby excluding themselves from such genetic input. In either case, Neanderthal fossils would present as being the same chronological age as the more modern looking *sapiens*, but would appear more "archaic" since they were geographically cut off from the continuing gene flow of the 'sons of gods.'

We might consider then that the European Neanderthal may be an earlier form of the so-called 'Black-headed Ones' which had less opportunity to commingle genetically with the space travelers since, at their first opportunity to do so, or perhaps because they were told to do so, they headed for the hills. Are the Neanderthals the Biblical patriarchs and their immediate descendants, who lived to be several hundred years of age?

There is obviously a lot of think about here, but it is clear that a worldwide deluge ultimately provides a clue to the enigma of the disappearing Neanderthals. Cuozzo posits that the Neanderthals survived the Great Flood and were in fact the Biblical patriarchs. Yet, according to *When the Earth Nearly Died*, the Neanderthals, for the most part, died in the Great Flood. Indeed, both of these statements could be true: some died and some survived. But, as Cuozzo points out, it is possible that the population of Neanderthals was fairly small to begin with. Perhaps those who survived the Great Flood became the Biblical patriarchs, who began the peopling of the world. They didn't populate the world single-handedly, however. According to the Sumerians, they had help from extraterrestrials who were compatible due to the genetic engineering of our species. Thus, outside genetic diffusion occurred in the descendent populations.

A Toolmaker by Any Name

In light of the findings in this chapter, I would also have to conclude as erroneous Sitchin's statement to the effect that the appearance of *Homo sapiens* occurred "virtually overnight," and further that there is no indication of "a gradual change from *Homo erectus*." Some of the reasons for these conclusions are the following:

- Louis B. Leakey concluded that australopithecines were a side branch and not in the direct lineage of *sapiens*. Since *Homo erectus* was thought to be a descendant of *Australopithecus*, Leakey also removed *erectus* from the line of human ancestry.
- Modern researchers have pronounced that the femurs of "Java Man" are not *Homo erectus*, but are the bones of fully modern *sapiens*, and are accepted as evidence for anatomically modern humans existing 800,000 years ago in Indonesia.

- It is quite possible there were creatures of fully modern type already at Olduvai Gorge in Africa during the earliest Pleistocene era.
- Several human-like femurs discovered in East Africa which were attributed to *Homo habilis* may have actually belonged to anatomically modern humans.
- There is strong evidence for the presence of rounded bola (sling stone) makers in Argentina approximately 3 million years ago.
- There is reasonable evidence to back the view that anatomically modern humans have co-existed with other primates for tens of millions of years.

In short, *Homo sapiens* is much older than the theory being peddled in museums all over the world. Furthermore, it can also be safely assumed that museums and text books are proliferating a great lie called "evolution."

With regard to the Extraterrestrial Hypothesis, however, the above information does not actually discount it. It only suggests that whatever we define as *Homo sapiens*, a somewhat advanced <u>toolmaking</u> hominid, may have been the creature upon whom the genetic alterations were performed, with *sapiens sapiens* being the result. The various creatures called the australopithecines do not need to be considered in this scenario at all, since these creatures were *not* toolmakers (no matter what you read in *Time* magazine dated 8/23/99). This *Time* article, entitled "Up From The Apes," states that earlier discoveries indicate "someone was using carefully manufactured stone tools" in the area of Gona, Ethiopia at about 2.5 million years B.P. The discovery of a "new" hominid type called *Australopithecus garhi*, hailed as possibly "the first hominid to use stone tools and eat meat," made the cover of the magazine. However, the article stated that the evidence pointing to *A. garhi* as the "gifted toolmaker" was "circumstantial." It also suggested that, since the materials the tools were made from were not local, "these hominids must have carried their tools with them."

This article has the audacity to state that australopithecines were toolmakers with the presence of mind to carry their tools around. It also asserts that australopithecines "walked upright" about 4.2 million years B.P., although it states that they didn't walk "in the modern sense." However, a graph of human evolution within these pages shows these creatures walking upright with a completely straight-backed modern gait. Of course, a picture speaks a thousand words, and this picture carries the clear import that australopithecines were human ancestors, contrary to strong arguments that they were not. This article constitutes an outright fabrication of the factual evidence.

Australopithecines were most likely extinct apes, and, if we consider for the moment the genetic engineering hypothesis, these creatures most likely would not have been considered adequate for the purposes of genetic engineering a slave race. As Sitchin has pointed out, Sumerian accounts describe a creature who was capable of letting other animals loose from traps, which implies intelligence. This was another creature entirely. And it was apparently around at the same time that australopithecines were, since "somebody" made these tools, and it was not an ape.

If the family of man, early *sapiens*, co-existed with these australopithecine types for even so much as 2-3 million years, where did *sapiens* come from? If Cremo and Thompson's "lunatic fringe" evidence is ascertained as true, there is no evolutionary scenario that could possibly be derived from this data. It effectively cuts off all of evolution's possible "missing links." The myth of toolmaking australopithecines must be paraded before us in *Time* magazine images and graphs so that people have the sense that this concocted scenario represents hard cold facts. But it does not.

So, we're back where we started. Who came up with "person man"? We would have to conclude either an earlier seeding or genetic engineering by ETI space travelers, or we have to conclude a "supernatural" origin: i.e. creation.

The Ancient Family of Man

Evidence for an even more remote origin of humans has also been uncovered, but is, of course, heavily disputed. The Macoupin County skeleton, for instance, was discovered in Illinois in December of 1862, in a coal layer dated 286-320 million years. Fossilized human-like footprints were found in Kentucky in 1938, in rock dated at least as old as 300 million years, i.e. the "Carboniferous Period." Since the currently accepted paradigm asserts there were not even mammals at this time, it has been suggested that the prints belonged to some sort of bipedal amphibian. The problem, Cremo and Thompson argue, is that "human-sized Carboniferous bipedal amphibians do not fit into the accepted scheme of evolution much better than Carboniferous human beings." Geologist Albert Ingalls has stated: "… unless the Sumerians had airplanes and radios and listened to Amos and Andy, these prints were not made by any Carboniferous Period man."

I don't know if they listened to Amos and Andy, but the Sumerian civilization seems to have somehow gotten around the globe pretty well! Other anomalous annoyances to the status quo evolutionary paradigm discussed by Cremo and Thompson are the following artifacts:

- A human figurine, densely coated with iron oxide, found in a 300-foot well in Idaho dated 2 million years.

- A mortar and pestle removed from a deep mine tunnel under Table Mountain, California, found in gold-bearing gravel dated 33-55 million years. Fossils of human skulls have also been found in these deep mine shafts. In addition, a stone

Hundreds of metallic "spheres" have been discovered in a South African mine shaft in a Pre-Cambrian mineral deposit, dated at 2.8 billion years.

pestle embedded in gold-bearing gravel dated over 9 million years is in the hands of the Smithsonian.

• Several semi-ovoid "metallic tubes" of identical shape and varying sizes discovered in France in a 65-million year old (Cretaceous) chalk bed.

• A wall of polished 12-inch cubes or blocks discovered in Oklahoma in a 2-mile-deep (Carboniferous) coal mine shaft dated at 286 million years. Other stories of massive walls found in deep mine shafts indicate that this is a potential area of research.

• A shoe-like print found in Cambrian shale in Utah dated at 505 million years.

• Hundreds of metallic "spheres" discovered in a South African mine shaft in a Pre-Cambrian mineral deposit, dated at 2.8 billion years. At least one of these extremely hard ancient "bowling balls" has three straight parallel grooves running around the entire circumference of its equator, as though it may represent the planet Saturn. There are two types of spheres: one type is a solid bluish metal with white flecks, and the other type is a hollow ball filled with a white spongy center. The spheres are housed in a museum in Klerksdorp, South Africa. The curator states: "They look man-made, yet at the time in Earth's history when they came to rest in this rock *no intelligent life existed.*"

Perhaps it is time we changed this assumption. Many have attempted to discredit these and a vast number of incredible finds over the years. The result has been, as Cremo and Thompson indicate, if the "facts" do not fit the prevailing evolutionary theory, "the facts must go."

The "Seeding" of Life

Sitchin writes that the genetic manipulations forced upon mankind would only have been possible if "man had developed from the same 'seed of life' as the Nefilim." He also notes that Ea, the chief scientist of the Nefilim, wondered if the celestial collision with Nibiru had seeded Earth with its life. This might suggest that early hominids were "akin to the Nefilim—though in a less evolved form." With reference to the relationship between this collision and life on Earth, Sitchin wonders, if life on Earth 'germinated' from life on the Twelfth Planet, then would evolution on Earth have proceeded in the same way that it had on the Twelfth Planet? Undoubtedly, he writes, there were "mutations, variations, accelerations, and retardations caused by different local conditions," but, he suggests, "the same genetic codes, and the same 'chemistry of life' found in all living plants and animals on Earth would also have *guided the development* of life forms on Earth *in the same general direction* as on the Twelfth Planet."

Yet, if Darwinian theory is to be strictly applied, Sitchin's assessment that, if life on Earth came from life on the "12th planet," evolution on Earth would be *guided* in *the same general direction* as it had on the "12th planet," is erroneous. First, do we know for a fact that intelligent life 'germinates' from inorganic matter? No, we do not. Second, rather than concurring with the *acciden - tal* notions of Darwinian theory, this view instead concurs with a *purposeful* vitalist or teleological model, which views the process of evolution as inwardly *guided* toward an end result.

While Sitchin's timetable seems to mesh with the Out of Africa model, there is reason to question whether the Out of Africa model accurately portrays the emergence of mankind. In addition, there is still the question of whether space travelers (the Anunnaki or another group) were here prior to the documented visit of approximately 450,000 years ago. According to the Sumerian story called *Kingship in Heaven*, indeed an earlier Nibiruan visit is documented, but the details of this visit are unknown. Problematically, when I asked Sitchin in a 1995 interview where the hominids on the planet originally came from, his matter-of-fact answer was: "evolution." He indicated that the Anunnaki were also a product of evolution on their own planet, although I do not know how he deduced this information. He indicated that Sumerian writings state that "life evolved on Earth from the seed brought into the solar system by Nibiry. It's the same seed of life we call DNA."

It is not clear whether this statement indicates that the DNA was carried here via an Earth *collision* with the planet Nibiru, or whether it was carried *by hand* by the beings who reside on the planet. In any case, it suggests that the "seed" did not evolve on this planet. I pressed for a better answer. I asserted: "then the early hominids still were not here by 'accident.' Sitchin responded: "If you want to say that the celestial collision which caused life on Earth was not an accident, but was pre-determined by whoever runs things in the Universe, then it was no accident." However, in light of information contained in this chapter, there is no reason to believe that man's arrival on this planet was due to an "accident" called evolution.

Could Sitchin's translation of the Sumerian word for "seed" be incorrect? As he notes, the original Sumerian texts describing the cloning process have not yet been found, but he surmises that it had something to do with the genetic compatibility of the sperm and the egg. A type of "seed" alluded to in ancient Vedic texts is called "bijas." This type of seed contains the subtle energy from which life forms are generated. The Vedic philosophy teaches that a hierarchy of creative beings, or archetypes, are responsible for the creation and dissemination of different life forms. The Devas of ancient Vedic texts are the archetypes for all living creatures on Earth. For humans to descend from the Devas, something else is needed—i.e., a "transformation" must take place in order to convert the subtle energy into "gross energy," or material substance. Therefore, the "seeds" of life may refer to a vibratory transference of subtle energy to material energy, rather than to the physical DNA structure. This would essentially be considered an act of special creation.

As NASA's "Astrobiology Roadmap" indicates, "determining the primary sources and nature of organic matter from which living systems emerged on the prebiotic Earth is still a controversial endeavor." One objective is to: "determine whether the atmosphere of the early Earth, hydrothermal systems or exogamous matter were significant sources of organic matter." They also indicate it has not actually been ascertained that "life from one world can establish an evolutionary trajectory on another," but it is considered one of their goals to understand the natural processes by which life can migrate from one world to another. In other words, can the "seed" of life grow into a civilization of intelligent humanoids on its own accord? Is it programmed to do that? Not only is this a great leap of faith, but it is anti-Darwinian, since it supposes a force, field, energy or entity as a guiding teleological factor.

It is evident from Velikovsky's work that a scenario involving numerous catastrophes, followed by extinctions and creations, more accurately describes the fossil record than does the evolution of various forms from one single-celled creature. As NASA's Astrobiology web site states:

> Throughout its history, life on Earth has experienced [environmental] changes with events ranging from impacts of asteroids and comets, and their resultant global manifestations, to ice ages of varying duration. Throughout each of these changes, life generally has responded initially with reductions in genetic diversity, followed by recoveries and continued increases in biodiversity.

However, it has also been noted by NASA scientists that not only has Earth been hit by asteroids and comets (resulting in "global manifestations"), but it was also hit by a planetary body "the size of Mars," which ripped it in half and gave us a Moon for a thank you! Could any biological life forms, besides insects perhaps, survive such a large-scale catastrophic event? Does this mean life had to evolve again and again many times on its own? But haven't we been told this is impossible? As the famous line in Oliver Stone's *JFK* goes, "people, we're through the looking glass here."

As Velikovsky has so succinctly put it, the "first" act of human creation is obscured by the catastrophes following it, therefore, it depends on the memory of the people which catastrophe they consider as the first act of human creation. As he wrote: "Human beings, rising from some catastrophe, bereft of memory of what had happened, regarded themselves as created from the dust of the earth." This statement is a potential *key* to many mysteries. We can hypothesize from this, and from the information in Cremo and Thompson's *Hidden History of the Human Race*, that the guided creation of all life forms is ongoing in the Universe, and that the human form, or something akin to the human form, has been created, or brought in from elsewhere, many times and, as well, destroyed many times. This implies "creation" in one of its many aspects: either the concept of God, the existence of some other vital principle, or one or more groups of space-faring intelligent races, perhaps utilizing remote robotic genetic collection devices, or "UFOs."

Actually, NASA is very interested in the latter, and is developing procedures that will mimic these actions in the future. In its Astrobiology Roadmap web site, NASA states that the "potential seeding of Earth life on other planetary surfaces, both intentionally and unintentionally, is possible." It also states:

> The current possibilities for major impact of human activities on the terrestrial biosphere constitute an excellent observational laboratory to test the vulnerability of ecosystems. … Studies of ecosystem response to rapid environmental changes will help extend ecosystem models on the Earth and to other worlds, allow predictions of responses to major, planet-wide changes, and identify limits to these changes beyond which life may not be able to recover.

There are obviously huge chunks of information missing from the written record, as well as other information which we may be misinterpreting at this time due to anthropocentric bias. Yet, certain Sumerian writings seem to indicate that ET groups have been engaged in the project of seeding the planets they visit with their own DNA, as well as the DNA of other life forms. For instance, as Sitchin explains, it was the DNA of each of Earth's animal species, not the animals themselves, which was secured a place on the vessel with Noah and family. Also, the seeds for all of the post-Flood vegetation and grains, as well as the DNA of all of the post-Flood animals, were provided by the Anunnaki/Nibiruans after the deluge. It is not more than a stone's throw to surmise that these traveling biochemical engineers, or other advanced beings like them, visited Earth and seeded the planet at earlier times or, perhaps even, at regular intervals. We might even suspect that the fossil record, which exhibits species arriving "all at once and fully formed," is a record of such exogamous "seedings," or creations, of life forms.

Who Came Up With Person-Man?

This intriguing creation/evolution tale remains problematical. We have still not uncovered the First Cause: whether it was God, a Gaian animate, archetypal or other vital principle, or ETI space travelers, which originated the life forms on planet Earth. Nor do we have any information with respect to the origin of life on Nibiru. Sitchin notes in *The Twelfth Planet*, that of the 30,000 texts found at Nippur in Sumer, many remain unstudied and others may be still unrecovered. These may hold the key.

Yet, it would appear that Sitchin's stance on this subject has changed in a later book, *Divine Encounters*, wherein he announces that "the biblical suggestion that the Elohim—the 'gods,' the Anunnaki—had a God, seems totally incredible at first, but quite logical on reflection." Sitchin maintains that the Bible itself provides the answer to the question, "who created the Nefilim, the

Anunnaki, on their planet?" He asserts that the Bible is keenly aware of the planet called Nibiru, and called it *Olam*, the "disappearing planet." The Dogon people are also aware of a planet they say is "invisible." The prophets spoke of *Olam* as a measure of an extremely long time, the span of time between the periodic disappearance and reappearance of the planet on its 3,600 (Earth) year orbit. The Bible states in Psalm 61:8 "Before the *Elohim* upon *Olam* He sat."

This verse, Sitchin maintains in *Divine Encounters*, provides the physical location where God the creator, "He who is first and last," resides. Sitchin asserts that "no less than eleven times, the Bible refers to Yahweh's abode, domain, and 'kingdom' using the term *Olamin*, the plural of *Olam*—a domain, an abode, a kingdom that encompasses many worlds." The Dogon people call this planet "the egg of the world," and they say it is the origin of all things. They assert that it is "the center of all things and without its movement no other star could hold its course." The Sumerians write that: "All else—celestial planetary 'gods,' Nibiru that remade our solar system and remakes the Earth on its near passages, the Anunnaki 'Elohim,' Mankind, nations, kings—all are His manifestations and His instruments, carrying out a divine and universal everlasting plan."

Clearly, Sitchin's belief in 'Darwinian' evolution does not mesh with this latest concession that there is a God, a creator with a divine plan for all of creation, and that, according to Sumerian texts, he physically resides on a planet within our solar system. Since Darwinian evolution is incompatible with ultimate purpose or design, I point out this discrepancy to show that those of us interested in finding a resolution to this problem have got to throw one of these incompatible theories out the window. In essence, the ability to create a hybrid capable of reproducing on its own is *acting in the capacity* of a divine and intelligent creator. Most Christians continue to refute the Extraterrestrial Hypothesis because they think it takes God out of the picture, but might this group of obviously superior humans have "permission" to act in the capacity of a divine and omniscient God? Might they be the angels—the Elohim—who "work" for God?

In light of recent advances made in cloning, how far away from this reality are we humans already? Is this tale so unbelievable? Is it less believable than the convoluted tale of the emergence of *Homo sapiens* by *mutation* and *natural selection*? Which story contains more elements of myth?

According to Arthur Horn in *Humanity's Extraterrestrial Origins*, "at present there is no alternative to the neo-Darwinian theory in Western societies, except the creationist story in the Book of Genesis." He writes, "most neo-Darwinists today react as if an attack on the current theory is an attack on evolution itself." Horn is hopeful that there will be a "viable alternative to neo-Darwinism that will explain our biological evolution, as well as taking into an account our spiritual existence." While he's waiting (instead of holding his breath or whistling Dixie), Horn looks at the theory of "junk DNA," which is, supposedly, "leftover DNA from earlier evolutionary history which is no longer needed." Apparently, the rearrangement of "junk DNA" takes place in the regulatory genes, and may be the key to "the jumps of evolution seen in the fossil record."

Problematically, the "fossil record" is a contrived notion in the first place and, further, it does not evidence a linear procession of changes indicative of Darwinian-style evolution, let alone any "jumps of evolution." What Horn means to say is that there is no alternative *acceptable* to neo-Darwinists, since the alternatives propose an outward causal force—or *elan vital*—which the Western scientific clique cannot quantify with any tools at their disposal. Observe how mankind's *toolness* forces him into so many material corners.

Much later in the book, Horn offers a short explanation of Rupert

Sheldrake's theory of morphogenetic fields, equating Carl Jung's collective unconscious (or "archetypes") with Sheldrake's proposed morphic fields, as a collective memory which guides the evolution of organisms of each species or variety. He suggests that humankind, having access to this vibratory and energetic record of thoughts and events (known also as the Universal Akashic Records) taps into this energetic field as a source of spiritual and consciousness evolution. Now Horn is on the right track!

I point out this example to illustrate one of many instances where we feel compelled to work within the boundaries of the acceptable paradigm, while keeping our alternative theories in the margins of our thoughts. We do not have to do this, and the Science of Experience—Phenomenology—which I will outline in my next book, is the slingshot in our back pocket. We do not have to play by the "house rules," because we own the Castle. The rules of phenomenology come before theirs. We claim an honorary position in the hierarchy. We claim squatter's rights, and the materialist statute of limitations is <u>over</u>.

The Vedic Account of Creation

Myths containing legendary personages were entrusted to initiates of secret orders throughout the ages and, many believe, these stories contained phenomenal information which only a few advanced people were able to fully comprehend. It is often said that the meaning of these stories are highly controlled planetary secrets. Our ability to comprehend these stories can only have shriveled pitifully since then. We can only go with gut feelings and instincts. If the sky people were not the planetary gods, they were certainly written into the play *at some point* as part of a cosmic hierarchy of beings which represented those gods.

These hierarchies of space, layers of time, and worlds within worlds, invisible to the naked eye, may bridge the metaphysical gap between those who believe the ancient 'gods'were planets and those who believe the ancient 'gods' were people. Uncomfortable in this metaphysical space, however, the modern mind demands closure. Some take the high road (the "scholarly" Freudian view) and others take the low road (the "sci-fi" alien view), and both end up mired in linguistic nuance, never the twain to meet. How can we bridge this linguistic gap? Like good phenomenologists, we must blur the edges.

According to ancient Vedic literature, various humanoid races are said to live in parallel or higher-dimensional realms encompassed within a larger world than that which we are able to perceive. These races live together cooperatively and are endowed with various "supernatural" abilities which are called, in ancient Vedic literature, *siddhis*. In his book, *Alien Identities*, Richard Thompson writes that the *siddhis* are "natural principles" or powers, and that it is possible that machines might be constructed which would take advantage of these natural principles.

The *siddhis* are powers which, in the expansive Vedic literature, many humanoid races are said to possess. Humans can potentially acquire *siddhis*, just a few of which include: mental telepathy, hearing and seeing at great distance, levitation, power to change the size of objects or living bodies, power to move objects from one place to another without traversing the space between, power to travel through physical objects, power of long distance hypnotic thought-control, invisibility and cloaking, power to assume various forms or generate illusory forms, power to enter and control another's body, and so on.

According to Thompson, in *Alien Identities*, Vedic literature also describes an ongoing natural process of evolution of consciousness in the Universe. In the larger scheme of things, souls acquire "higher and higher types of bodies" as they move to higher realms of consciousness. At the height of consciousness evolution is the state of complete liberation from the material world, involving "transfer of the soul to a completely transcendental realm." Vedic philosophy teaches that all manifestations on all planes of existence emanate from the Supreme Being. All souls share the qualities of the Supreme to a minuscule degree, but those encased in material bodies tend to have "perverted qualities due to the influence of the material energy."

According to the Vedic worldview, there is a Cosmic Hierarchy which consists of a "graded series of higher planetary systems, each of which is inaccessible to the inhabitants of the systems below it." The highest authority in the material Universe, Brahma, resides in the highest planetary system, Brahmaloka. Other planetary systems are inhabited by "sages" and "ascetics" who cultivate transcendental consciousness. Beneath these are the "Devas," who are organized in a type of military hierarchy. The Devas can take on any form they wish.

According to the ancient Vedic scripts, the Universe is under intelligent control. High-level controllers—the Devas and sages—do not generally intervene in human affairs. Rather, they arrange for the "transmigration" of souls from one material body to the next, in a process of movement toward higher consciousness. Their appointed cause is to guide embodied souls by the dissemination of spiritual teachings toward this higher ground. The transmigration toward higher consciousness also necessitates the change from gross energy to subtle energy. However, I think there is a mistake being made in UFO/New Age circles, to the effect that this higher-vibrational evolution is going to occur while we are in the living state. I believe this happens at death, and humans aren't going to start vibrating out of our skin en masse in the year 2012, or anything of the sort.

Among the many humanoid types, there are those dedicated to the service of the Supreme Being and others who are self-centered, being "greatly attracted to the exploitation of mystic powers and technology." All of these various groups are under the control of the universal hierarchy and, "are not able to act fully according to their own propensities." Some cosmic rebels, however, do "sometimes interfere strongly in earthly affairs." The Devas are of a "godly nature" and are said to hold "administrative posts" in the universal hierarchy. Greek and Roman mythologies refer to them as "demigods." Their close relatives, the Asuras, are the cosmic rebels who are often referred to as "demons." However, as Thompson notes in *Alien Identities,* the word demon comes from the word "Daemon," which the Romans and Greeks considered "intermediaries." These beings were not all regarded as evil until the later influence of Christianity.

The bodies of Brahma and the Devas, which emanate from the Supreme Being, are made of "subtle forms of energy," and do not contain DNA. But, they do "carry genetic information in the form of *bijas*, or seeds, that are also made of subtle energy." For humans to descend from the Devas, something else is needed—i.e., a "transformation" must take place in order to convert the subtle energy into "gross energy," or material substance. It is evident here that there may be a problem with Sitchin's translation of "seed" as "DNA," and this may be the missing "substance" which his analysis seeks. The genetic engineering hypothesis may be a mistaken translation of a more subtle process or transformational principle. As Thompson maintains, the Vedic version of the origin of humankind on Earth is that they were not "genetically engineered" by crossing the DNA of the gods with the DNA of the primitive "ape-man," but that it "involved mating between Devas that generated human offspring through preplanned genetic transformations."

Vedic texts teach that all the species of living beings were brought into existence by "a process of creation and emanation." It is taught that Brahma generated bodily forms by "direct mental action," and that the generations of descendants were produced from these forms by sexual reproduction. However, these descendants carried *bijas* for many different types of beings, and could produce different types of offspring. Vedic texts teach that different humanoid races on separate worlds were produced in this manner and, thus, have a common ancestry and are genetically compatible. Vedic texts suggest the humanoid form is actually common in the Universe. As Thompson writes (p. 295): "All of the humanoid races descend from male and female forms generated by Brahma, the original created being. The Devas are among the descendants of these forms, and earthly human beings are descended from Devas along a number of different lines of ancestry."

The Devas, along with other beings on that level or higher, contained the "seed," or programmed capacity on the creative level, to be "able to produce offspring that were not of their own bodily type." This seed (*bijas*) is not to be confused with DNA, as it is a type of subtle energy information storage. These types of beings are able to generate any form of living being by a combined process of mating and transformation of subtle to gross energy. The conversion of genetic information from subtle form to gross form is simply a matter of information storage utilizing different types of energy. The Devas can be thought of as "archetypes," or as giving birth to the archetypes for all living creatures on the Earth.

On the subject of archetypes, Max Heindel writes in *The Rosicrucian Cosmo-Conception*, (p.50):

> When we speak of the archetypes of all the different forms in the dense world it must not be thought that these archetypes are merely models in the same sense in which we speak of an object constructed in miniature, or in some material other than that appropriate for its proper and final use. They are not merely likenesses nor models of the forms we see about us, but are creative archetypes; that is, they fashion the forms of the Physical World in their own likeness(es), for often many work together to form one certain species, each archetype giving part of itself to build the required form. ... They plan changes as an architect plans the alteration of a building before the workmen give it concrete expression.

Thompson points out the same problem with Sitchin's theory that I have been wrestling with for quite some time. In Sitchin's theory, he writes, the ETs presumably *evolved* on another planet. However, he asks, why would such ETs be close enough genetically to Earth species to make this crossbreeding possible? Problematically, it has been written that the Anunnaki were actually of the "lizard" race. How could they have combined their own genes with those of an Earth-based mammalian/humanoid form?

Thompson points out that the Vedic account avoids the genetic incompatibility problem by "starting with the Devas and positing a transformation that alters the Deva form." The resulting human form is different from the Deva form, but close enough to make cross-breeding between humans and Devas possible. This may explain Sitchin's mistranslation of the "second trial" of the genetic engineering hypothesis.

However, Thompson writes, there are also Vedic accounts of the creation of human races via processes which sound similar to cloning, and both types of creation have been noted in Vedic literature. One way to get around these issues is to presume that both have occurred, but on different levels. The resolution of these issues lies in the hierarchy of beings described in Vedic literature. This hierarchy is also described in the Sumerian literature, and Sitchin relays the fact that the Anunnaki were in contact with beings higher than themselves.

According to the Vedic texts, the Devas (and others in that level and above) do not need to manipulate DNA in laboratories; they only need to give birth to the archetypes of all forms of living beings which are inherent in them as manifestations of the Supreme Being. Their work on all planetary systems is ongoing, and they have performed their work on the planet Earth, perhaps many times, particularly following mass extinctions.

I would suggest that the ETs who settled in Sumeria were below the Devas in this hierarchy, and therefore *did* require the laboratories, paraphernalia, and know-how to perform cross-breeding experiments to obtain a new species. This is "creation" on a totally different level: the genetic level.

Were We Cloned?

This book has explored the idea that we did not evolve from the great ape lineage. If we didn't evolve from the ponds of our beautiful blue oasis, where did we come from?

There are two choices: divine creation by an omniscient power which may be hierarchical in nature (i.e. the term *nature* meaning "by definition" as well as "discoverable scientifically"), or genetic engineering by cosmic hoodlums, with or without the explicit permission of such creative natural powers. Sitchin's ET-Seeding hypothesis asserts that we came into existence simply to be put to work and, seemingly, to be annihilated when the work was done. Who in their right mind wants to hear this?

After reading this manuscript, a Christian friend commented that it is a "construct." Yes it is. It is a construct of a construct of a construct, ad infinitum. Yet, this has nothing to do with any concepts of truth or falsity. All philosophical ideas are shared personal constructs. Consensus doesn't automatically make them true. As libertarian ethics teaches us, a lone personal construct is no less true than a shared one. Agreement is the cement of social structure. But at the bottom of it all is a construct, like it or not.

We like to think of ourselves as creatures born into a meaningful Universe. On the face of it, the Universe being offered for consideration here is patently absurd. It's straight out of the Twilight Zone. But is it really more absurd than the one we've got now? We flock to our churches to find an ideological meaning for our absurd reality. And the answer we find there is that there is a "special plan" for us that, in our infinite stupidity, we cannot possibly comprehend. I personally can't get close to that answer.

It is understandable that the idea of being "genetically engineered" by humanoids from another planet elicits an emotional tone with most people. We thought we were "special" creations, with a special plan in God's mind. However, the idea of a hierarchy of archetypal entities, the Elohim or the Devas, who assist in the ongoing creation of life forms, is an idea I can get close to, and I believe it is compatible with religious doctrines.

As we get a little more comfortable with the idea of the Earth as a controlled DNA repository for the ongoing special creations of the Devas, we will eventually have to consider the possibility of "fresh creations" of life at intervals throughout Earth history. This will be a time for correcting the textbooks. But this will have to await another Age.

Resources and Recommended Reading

Allen, D.S. and J.B. Delair, *When the Earth Nearly Died: Compelling Evidence of a Catastrophic World Change 9,500 B.C.*, Gateway Books, Bath, England, 1995.

Allen & Delair, *Cataclysm: Compelling Evidence of a Cosmic Catastrophe in 9500 B.C.*, Bear & Co., 1997.

The Anomalist, P.O. Box 577, Jefferson Valley, NY 10535.

Barrow, John D. and Frank J. Tipler, *The Anthropic Cosmological Principle,* Oxford Press, 1988.

Behe, Michael J., *Darwin's Black Box: The Biochemical Challenge to Evolution*, Simon & Schuster, New York, 1996.

Blavatsky, Helena, *Isis Unveiled, Secrets of the Ancient Wisdom Tradition*, Theosophical Publishing House, IL, 1997.

Peter Bowler, *The Non-Darwinian Revolution: Reinterpreting a Historical Myth*, Johns Hopkins Press, 1992.

Bramley, William, *The Gods of Eden,* Avon, NY, 1993.

Capra, Fritjof, *The Web of Life: A New Scientific Understanding of Living Systems* , Anchor Books/Doubleday, 1996.

Carlotto, Mark, *The Martian Enigmas: A Closer Look*, North Atlantic, 1997.

Coffey, E.J., "The Anthropomorphic Fallacy," *Journal of the British Interplanetary Society* 1992.

Corso, Philip, *The Day After Roswell*, New York, Simon & Schuster, 1997.

Cremo, Michael, Richard Thompson, *The Hidden History of the Human Race*, Bhaktivedanta Publishing, Los Angeles, 1999.

Cuozzo, Jack, *Buried Alive: Startling Truth About Neanderthal Man,* Master Books, 1998.

d'Arc, Joan (Ed.), *Paranoid Women Collect Their Thoughts*, Paranoia Publishing, 1996.

Davies, Paul, The Mind of God: The Scientific Basis for a Rational World, Simon Schuster 1992.

De Santillana, George, and Hertha von Dechend, *Hamlet's Mill: An Essay Investigating the Origins of Human Knowledge and Its Transmission Through Myth*, Godine, NH, 1999.

Ditfurth, H. V., *The Origins of Life: Evolution as Creation*, Harper & Row.

Fitzgerald, Randall, *Cosmic Test Tube: Extraterrestrial Contact, Theories and Evidence*, Moon Lake Media, 1998.

Fort, Charles, *The Book of the Damned*, Prometheus, reprint 1999.

Freer, Neil, "From Godspell to God Games," in *Of Heaven and Earth*, The Book Tree, CA, 1996.

Friedman, Stanton, *Top Secret/Majic*, Marlowe & Co., New York, 1996.

Frissell, Bob, *Nothing in This Book is True, But its Exactly How Things Are: The Esoteric Meaning of the Monuments of Mars*, No. Atlantic Books, 1994.

Goswami, Amit, *The Self-Aware Universe: How Consciousness Creates the Material World*, Tarcher, New York, 1995.

Hamilton, William F., *Alien Magic*, Global Communications, 1996.

Hancock, Graham, *The Mars Mystery: Secret Connection Between Earth and the Red Planet*, Three Rivers Press, 1999.

Herbert, Nick, *Elemental Mind: Human Consciousness and the New Physics,* Penguin, NY, 1993.

Hoagland, Richard, The Mars/Moon Connection (Video), Evidence of a Solar System Filled with Ancient Artifacts (Video)

Hoagland, Richard, *Monuments of Mars: A City on the Edge of Forever*, North Atlantic Books, 1996.

Hoagland & Bara, www.enterprisemission.com/see also www.lunaranomalies.com

Hoffman, Michael, III, *Secret Societies and Psychological Warfare.*

Horn, Arthur, *Humanity's Extraterrestrial Origins: ET Influences on Humankind's Biological and Cultural Evolution*, AZ, 1996.

Johnson, Charles William, *Earth/matriX: Science in Ancient Artwork*, www.earthmatrix.com.

Johnson, Phillip, *Darwin on Trial*, InterVarsity Press, Downer's Grove, IL.

Jones, Scott, (Ed.) *When Cosmic Cultures Meet*, Human Potential Foundation, 1995.

Jung, Carl, *Flying Saucers: A Modern Myth of Things Seen in the Skies.*

Key, Wilson Bryan, *The Age of Manipulation: The Con in Confidence, The Sin in Sincere*, Madison Books, Lanham, MD, 1989, 1993.

Kuhn, Thomas S., *The Structure of Scientific Revolutions*, (3rd.) Chicago, 1996.

Lasker and Tyzzer, *Physical Anthropology*, CBS College Publishing, 3rd ed., 1982.

Leakey, Richard, *The Origin of Humankind*, Harper, New York, 1994.

Mann, Nicolas & Marcia Sutton, *Giants of Gaia*, Brotherhood of Life Books, 1995.

Marrs, Jim, *Alien Agenda: Investigating the Extraterrestrial Presence Among Us*, HarperCollins, NY, 1997.

Martin, Graham, *Shadows in the Cave: Mapping the Conscious Universe.*

McDaniel, Stanley, *The Case for the Face: Scientists Examine the Evidence for Alien Artifacts on Mars*, Adventures Unlimited, 1998.

Messinger, Lanny, *The Programming of a Planet*, self-published. (also see "The Meaning of Light: A Synopsis of Extra-Terrestrial Intentions and Methods, *Paranoia: The Conspiracy Reader*, Spring, 1999, P.O. Box 1041, Providence, RI 02901)

Midgely, Mary, *Evolution as a Religion*, Methuen & Co., London/New York, 1985.

Milton, Richard, *Shattering the Myths of Darwinism*, Park Street Press, VT, 1997.

Pinkham, Mark Amaru, *The Return of the Serpents of Wisdom*, Adventures Unlimited, 1997.

Randle, Kevin, *The Truth About the UFO Crash at Roswell.*

Rene, Ralph, *Nasa Mooned America*, 31 Burgess Place, Passaic, NJ 07055.

Rux, Bruce, *Architects of the Underworld: Unriddling Atlantis, Anomalies of Mars, and the Mystery of the Sphinx*, Frog Ltd./North Atlantic, Berkeley, 1996.

Shklovskii & Sagan, *Intelligent Life in the Universe*, O/P.

Sheldrake, Rupert, *A New Science of Life: The Hypothesis of Morphic Resonance, Park Street Press,* VT, 1995.

Sitchin, Zecharia:

The Cosmic Code, Avon, NY, 1998.

Divine Encounters: A Guide to Visions, Angels and Other Emissaries, Avon, NY, 1996.

Genesis Revisited: Is Modern Science Catching Up With Ancient Knowledge, Avon, NY, 1995.

The Twelfth Planet, Avon, NY, 1983.

The Wars of Gods and Men, Avon, NY, 1999.

Society for Interdisciplinary Studies, (information on Catastrophism, Archeoastronomy, etc. at www.knowledge.co.uk/sis)

Swann, Ingo, *Penetration: The Question of ET and Human Telepathy*, TwiggsCompany, SD, 1998. (www.biomindsuperpowers.com)

Tattersal, Ian, *The Last Neanderthal: The Rise, Success and Mysterious Extinction of Our Closest Human Relatives*, Simon & Schuster, NY, 1995.

Temple, Robert K.G., *The Sirius Mystery*, Destiny Books, VT, 1987.

Thompson, Keith, *Angels and Aliens: UFOs and the Mythic Imagination*, Ballantine, NY, 1991.

Thompson, Richard, *Alien Identities: Ancient Insights into Modern UFO Phenomena.*

Velikovsky, Immanuel, *In the Beginning,* (www.velikovsky.collision.org)

Index

TRIUMPH OF THE HUMAN SPIRIT: The Greatest Achievements of the Human Soul and How Its Power Can Change Your Life by **Paul Tice.** A triumph of the human spirit happens when we know we are right about something, put our heart into achieving its goal, and then succeed. There is no better feeling. People throughout history have triumphed while fighting for the highest ideal of all – spiritual truth. Some of these people and movements failed, other times they changed the course of history. Those who failed only did so on a physical level, when they were eliminated through violence. Their spirit lives on. This book not only documents the history of spiritual giants, it shows you how you can achieve your own spiritual triumph. Various exercises will strengthen your soul and reveal its hidden power. In today's world we are free to explore the truth without fear of being tortured or executed. As a result, the rewards are great. You will discover your true spiritual power with this work and will be able to tap into it. This is the perfect book for all those who believe in spiritual freedom and have a passion for the truth. **(1999) • 295 pages • 6 x 9 • trade paperback • $19.95 • ISBN 1-885395-57-4**

PAST SHOCK: The Origin of Religion and Its Impact on the Human Soul by **Jack Barranger. Introduction by Paul Tice.** Twenty years ago, Alvin Toffler coined the term "future shock" – a syndrome in which people are overwhelmed by the future. *Past Shock* suggests that events which happened thousands of years ago very strongly impact humanity today. This book reveals incredible observations on our inherited "slave chip" programming and how w've been conditioned to remain spiritually ignorant. Barranger exposes what he calls the "pretender gods," advanced beings who were not divine, but had advanced knowledge of scientific principles which included genetic engineering. Our advanced science of today has unraveled their secrets, and people like Barranger have the knowledge and courage to expose exactly how we were manipulated. Learn about our past conditioning and how to conquer the "slave chip" mentality to begin living life as a spiritually fulfilled being. **(1998) • 126 pages • 6x9 • trade paperback • $12.95 • ISBN 1-885395-08-6**

GOD GAMES: What Do You Do Forever? by **Neil Freer. Introduction by Zecharia Sitchin.** This new book by the author of Breaking the Godspell clearly outlines the entire human evolutionary scenario. While Sitchin has delineated what happened to humankind in the remote past based on ancient texts, Freer outlines the implications for the future. We are all creating the next step we need to take as we evolve from a genetically engineered species into something far beyond what we could ever imagine. We can now play our own "god games." We are convinced that great thinkers in the future will look back on this book, in particular, as being the one which opened the door to a new paradigm now developing. Neil Freer is a brilliant philosopher who recognizes the complete picture today, and is far ahead of all others who wonder what really makes us tick, and where it is that we are going. This book will make readers think in new and different ways. **(1998) • 310 pages • 6 x 9 • trade paperback • $19.95 • ISBN 1-885395-39-6**

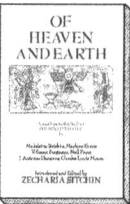

OF HEAVEN AND EARTH: Essays Presented at the First Sitchin Studies Day. **Edited by Zecharia Sitchin.** Zecharia Sitchin's previous books have sold millions around the world. This book contains further information on his incredible theories about the origins of mankind and the intervention by intelligences beyond the Earth. This book offers the complete proceedings of the first Sitchin Studies Day. Sitchin's keynote address opens the book, followed by six other prominent speakers whose work has been influenced by Sitchin. The other contributors include two university professors, a clergyman, a UFO expert, a philosopher, and a novelist – who joined Zecharia Sitchin to describe how his findings and conclusions have affected what they teach and preach. They all seem to agree that the myths of ancient peoples were actual events as opposed to being figments of imaginations. Another point of agreement is in Sitchin's work being the early part of a new paradigm – one that is already beginning to shake the very foundations of religion, archaeology and our society in general. **(1996) • 164 pages • 5 1/2 x 8 1/2 • trade paperback • $14.95 • ISBN 1-885395-17-5**

FLYING SERPENTS AND DRAGONS: The Story of Mankind's Reptilian Past, By **R.A. Boulay.** Revised and expanded edition. This highly original work deals a shattering blow to all our preconceived notions about our past and human origins. Worldwide legends refer to giant flying lizards and dragons which came to this planet and founded the ancient civilizations of Mesopotamia, Egypt, India and China. Who were these reptilian creatures? This book provides the answers to many of the riddles of history such as what was the real reason for man's creation, why did Adam lose his chance at immortality in the Garden of Eden, who were the Nefilim who descended from heaven and mated with human women, why the serpent take such a bum rap in history, why didn't Adam and Eve wear clothes in Eden, what were the "crystals" or "stones" that the ancient gods fought over, why did the ancient Sumerians call their major gods USHUMGAL, which means literally "great fiery, flying serpent," what was the role of the gigantic stone platform at Baalbek, and what were the "boats of heaven" in ancient Egypt and the "sky chariots" of the Bible? **(1997, 1999) • 276 pages • 6 x 9 • trade paperback • $19.95 • ISBN 1-885395-38-8**

www.ingramcontent.com/pod-product-compliance
Lightning Source LLC
Chambersburg PA
CBHW022056210326
41519CB00054B/479